Praise for *Healthy Bee, Sick Bee*

'If you want to learn about popping bee penises (and honestly, who doesn't), and where not to get stung, then this is the book to read. . . . I can easily see *Healthy Bee, Sick Bee* becoming the go-to-book for anyone interested in honeybees and their demons, large and small. Highly recommended.'
—Prof Madeleine Beekman, School of Life and Environmental Sciences, University of Sydney

'Fascinating and informative . . . a very readable, enlightening text, interwoven with delightful anecdotes and personal touches.' —Judith de Wilde, *NZ Beekeeper*

Praise for *The Vulgar Wasp*

'Lester cleverly weaves facts and figures on the astonishing science of this little-loved insect into a text that's tickled with memorable anecdotes and personable insights. If you thought wasps were pointless, boring and unimportant, think again.'
—Dr Seirian Sumner, Reader in Behavioural Ecology at University College London

'*The Vulgar Wasp* is more than just interesting and instructive; it's a delight to read.'
—Dr Andrea Byrom, Director of New Zealand's Biological Heritage National Science Challenge

# PESTS AND PESTILENCE

The management of invasive species,
pests and disease in New Zealand

## PHIL LESTER

TE HERENGA WAKA
UNIVERSITY PRESS

Te Herenga Waka University Press
PO Box 600 Wellington
New Zealand
teherengawakapress.co.nz

A catalogue record for this book is available from
the National Library of New Zealand.

ISBN 9781776920082


Printed in Singapore by Markono Print Media Pte Ltd

# CONTENTS

# Preface

I have some bad news. We are very likely to have a future in which our problems with pests, invasive species, pestilence and plagues are only going to get worse. As you'll see in this book, that prediction isn't just mine; it is an evidence-based and widely held conclusion.

Human management of pests and diseases has a long, rich and fascinating history. In the future, our management of pests and pestilence will be especially important, since the world's population is likely to reach nearly 10 billion people by 2050. We will need approaches that are effective and socially accepted, and that cause the least possible non-target effects.

My initial purpose in writing this book was to provide a resource for a course I teach at Te Herenga Waka—Victoria University of Wellington on pests and pest management. I wanted to provide an accessible account that would help readers understand the reasons and mechanisms for species to become pests or plagues, and what we are doing or could do about them. But I also wanted to make this work comprehensible and useful to a wider audience. I have included stories of our history with pests and plagues. There are cats on parachutes, a potato famine that led to Irish immigrants living all over the world, and an asymptomatic but superspreading cook who gave typhoid to at least 51 other people.

Our road ahead is paved with the potholes of more pests and more pestilence. We, our children, and all future generations will have no choice but to deal with these issues, along with our legacy of habitat destruction and climate change. I hope we can act today to help reduce the toll of pests and pestilence on the world in which we live. I hope that this book contributes in some way to these goals. Personally, I think this is also a captivating topic.

*An Irish Peasant Family Discovering the Blight of Their Store*, a painting by Irish artist Daniel Macdonald depicting a family in 1847, the worst year of the Irish Potato Famine. Potato blight is caused by the microorganism *Phytophthora infestans*. *National Folklore Collection, University College Dublin / Wikimedia Commons*

# 1. THE WORLD'S PESTS

Those species that hurt economies, health or biodiversity –
or are just a nuisance

The Great Chinese Famine is regarded as the deadliest in human history. This human-made disaster caused more than 30 million people to starve to death over the period of 1959–61. There are reports of desperation so extreme that people resorted to cannibalism. Human flesh was sold in markets, and children were swapped between families so that people could consume them as food without committing the sin of eating their own.[1]

What caused this famine? The Chinese Communist Party and Mao Zedong had swept into power in 1949. They immediately implemented a programme to improve public health. Their targets included a long list of diseases and parasites afflicting their people, such as tuberculosis, the bubonic plague, polio, cholera, malaria, smallpox and hookworm. Infant mortality rates at the time were as high as 30%.[2] Water sanitation and vaccination programmes were successful in lowering the rates of many of these diseases. The animals that transmitted the diseases and others that consumed crops also became targets. A 'Four Pests' campaign challenged citizens to 'eradicate pests and diseases and build happiness for ten thousand generations'.[2] The chosen pests were sparrows that eat rice and grain; rats that consume food stores and host disease-transmitting fleas; malaria-spreading mosquitoes; and those pesky flies.

In 1958, the slaughter began. The Chinese government claimed to have killed 1 billion sparrows, 1.5 billion rats, 100 million kilograms of flies and 11 million kilograms of mosquitoes.[3] The killing of sparrows had the greatest unanticipated effect. Their role in controlling ravenous crop-eating insects had not been appreciated. In the birds' absence, populations of locusts and other insect pests grew unconstrained and consumed crops unchecked. Crops failed and people starved. Other policies and practices added fuel to the famine flame, but the Four Pests campaign was a major contributing factor to the deaths of an estimated 30 million people. It is an extraordinary story of how animals can cause us harm, and how we humans can upset and damage an ecosystem, to our own detriment.

We all think we know what a pest is, but a great place to start this book is with a definition. A pest is generally defined as any animal or plant species that is detrimental to humans or human concerns, particularly creatures that substantially damage crops, forestry or livestock, or that are simply a nuisance to people. The locusts that ate Chinese crops and mosquitoes that spread malaria are great examples. This is a very anthropocentric definition. And, as we will see later, some people might view one species as a pest while others see a treasure. Trout and salmonids introduced for recreational fishing, Polynesian or Pacific rats, feral pigs in Hawai'i and even honey bees fall into these dual categories.

Pests to most people are animals or plants. Bacteria, microscopic fungi, protists and viruses are also detrimental to humans or human concerns, but are microscopic. Because they are tiny, these microorganisms are generally considered 'pathogens', or causes of disease, rather than 'pests'. Nevertheless, they are still detrimental to humans or human concerns. Pathogens and pests are

'Exterminate the four pests!' A 1958 poster encourages the Chinese people to 'eradicate pests and diseases and build happiness for ten thousand generations'. On the right, people display their sparrow kills on a cart during the Four Pests campaign in April 1958. *Left: Wikimedia Commons. Right: International Institute of Social History / Stefan R. Landsberger Collections*

often intertwined. Many animal and plant diseases require an arthropod vector. Mosquitoes and malaria, or the tsetse fly and the trypanosome parasite that causes sleeping sickness, are examples where the pathogen is entirely dependent on the pest for transmission.

In the Bible and other classical texts, an abundance of these microscopic pathogens and plant and animal pests are referred to as pestilence. One of the Four Horsemen of the Apocalypse is sometimes called Pestilence: 'They were given power over a fourth of the earth to kill by sword, famine, plague, and by the wild beasts of the earth' (Revelation 6:8). Pestilence is often depicted in barbarous attire riding a pale or white horse. In other media, Pestilence is a supervillain in comic books, and a death metal band in the Netherlands.

Given the definition's emphasis on 'detrimental to humans or human concerns', I'll discuss pests as animals, plants and pathogens in this chapter. As we'll see, Pestilence frequently does take his fourth of the earth.

## Pathogens and pests that are directly detrimental to humans

Which species causes the greatest global burden on people? Every country and part of the globe is afflicted with a different subset of pathogens and pests, and their influence varies over time. For example, each year people living in the United States have a real possibility of catching bubonic plague. Prairie dogs, squirrels and many other animals host this disease, which is estimated to have killed up to a third of all humans in Europe in the 14th century. People in Arizona, California, Colorado and many other states today can be exposed to the bubonic plague when pet cats pick up and bring home fleas carrying the bacterium. In recent times, the highest rates of bubonic plague, caused by the bacterium *Yersinia pestis*, occurred in Madagascar, where an outbreak in 2014 caused 263 human cases with a 27% fatality rate.[4] In contrast, bubonic plague isn't an issue if you live in Western Africa, but this region experienced the worst-known outbreak of the viral disease Ebola over 2013–16. Many more people were infected and died in this outbreak than in all previous Ebola outbreaks combined. At least 28,616 people contracted Ebola during this outbreak, with at least 11,000 deaths.[5]

Let's look at the species responsible for the sum total global burden of disease on human mortality (Fig. 1.1).

Historically, tuberculosis has been by far the biggest killer. A billion human lives have been taken by the bacterium *Mycobacterium tuberculosis* over the last 200 years.[6] The World Health Organization (WHO) estimates that a quarter of the world currently carries a latent infection of tuberculosis, with no obvious symptoms. Around a tenth of those people live in countries such as India, China, Nigeria and South Africa, which bear a high burden of the global tuberculosis cases. Famous people who have died or suffered from tuberculosis include the Polish composer Frédéric Chopin, the writer George Orwell and the communist Vietnamese revolutionary Hồ Chí Minh. The archbishop and anti-apartheid campaigner Desmond Tutu suffered from tuberculosis in his youth. Another of the mortally afflicted was the poet John Keats, who upon coughing up blood in

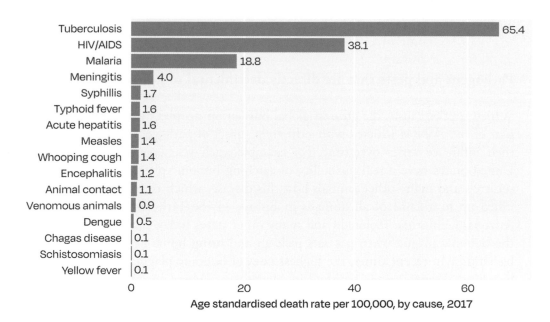

Fig. 1.1: Estimated rates of human mortality due to pathogens and animal pests. Many different pathogens cause human mortality, with the top 10 shown here in addition to other key parasites or pathogens. The rates for tuberculosis, HIV/AIDS and malaria incorporate co-infections such as malaria in combination with other tropical illnesses. Mortality directly from animals accounts for only two deaths per 100,000 people. The estimated total of all deaths from pathogens and pest amounts to 19.2% of the world's annual mortality. *Data from Roth et al. (2018).*[7]

1818 is widely quoted to have said, 'I know the colour of that blood; it is arterial blood; – I cannot be deceived in that colour; that drop of blood is my death-warrant – I must die.'[8] He had seen both his mother and brother succumb to tuberculosis. One in four Londoners died from tuberculosis in the early 19th century. Even today, London is seen as the tuberculosis capital of Europe, with parts of London experiencing higher rates of infection than Eritrea, Iraq and Rwanda.[9]

The second-most lethal pathogen afflicting human health over the last several decades is thought to be HIV/AIDS (Human Immunodeficiency Virus, causing the Acquired Immune Deficiency Syndrome). Monkeys and apes in Western Africa frequently carry simian immunodeficiency viruses (SIVs). Of all these SIVs, the closest relative to the globally widespread HIV strain (called HIV-1) has been isolated from wild chimpanzees (*Pan troglodytes troglodytes*). There are two other strains of HIV-1, indicating that at least three independent introductions of SIVs have come into the human population.[10] How the chimpanzee virus made the leap to humans is not known for certain, but this jump is likely to

**TUBERCULOSIS AND CHILDHOOD**

The State must protect its children.
Almost 50,000 American children die yearly of tuberculosis. The anti-tuberculosis campaign must concentrate on the child.

Almost all children have dormant tuberculosis germs in their bodies. To keep these dormant germs from developing active disease, train children in the proper use of

**FRESH AIR - SUNSHINE GOOD FOOD - REST**

© NATIONAL CHILD WELFARE ASSOCIATION, NEW YORK,
IN CO-OPERATION WITH
NATL. ASSN. FOR THE STUDY AND PREVENTION OF TUBERCULOSIS

Tuberculosis was a widespread cause of death before the use of a vaccine (from 1921) and antibiotics (from 1944). This poster from the 1920s, advocating fresh air, sunshine, good food and rest, was distributed in the US by the National Child Welfare Association and the National Association for the Study and Prevention of Tuberculosis. It was part of a set produced for a 'Modern Health Crusade' exhibition. The illustration shows Columbia – a quasi-mythological figure representing the spirit of the US – with children. *Simon Stone / Alamy*

have resulted from the unsafe preparation and consumption of chimpanzees as bushmeat. Genetic tracing has allowed scientists to home in on chimpanzee populations from southern Cameroon as the likely source of the infection.[11]

The virus that we know as HIV has been traced to the city of Kinshasa in what is now the Democratic Republic of Congo, with infections beginning in the 1920s.[12] Kinshasa (then called Leopoldville) now has the unfortunate label of 'the cradle of the AIDS pandemic'. For several decades, the predominant genotype of the virus responsible for the global HIV/AIDs pandemic remained only in the Democratic Republic of Congo and neighbouring countries, seemingly having spread along roads, railways and rivers via migrants and workers in the sex trade. The virus then spread from Africa to Haiti and the Caribbean in the 1960s. It travelled from the Caribbean to New York City around 1970, then to San Francisco later that decade, and then around the world. The arrival of HIV in the United States was signalled by doctors noticing patients suffering from pathogens that normally do not cause illness. Diseases such as pneumonia caused by the normally harmless parasitic fungal species *Pneumocystis jirovecii*, which you and

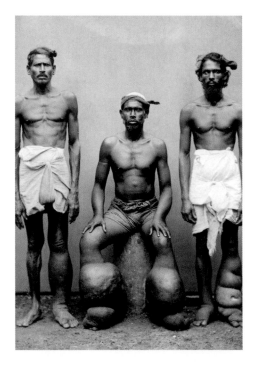

Three men suffering from lymphatic filariasis or elephantiasis. The disease is caused by a nematode parasite (*Onchocerca volvulus*) that is transmitted by blackflies of the genus *Simulium*. This photograph was taken c.1890 in South East Asia. The WHO aimed to eradicate the disease as a public health problem by 2020. Researchers have considered control programmes targeting the parasite or its insect pest vector, much like the approach to malaria.
*Pump Park Vintage Photography / Alamy*

I probably carry, became frequent and lethal amongst immunocompromised sufferers of HIV. A Californian doctor in 1980 described a patient afflicted by this disease as pale, extremely thin and ashen, bordering on anorexic, coughing painfully and uncontrollably, with a mouth full of a white 'cottage cheese' infection.[13] This form and cause of pneumonia is still a frequent AIDS-defining diagnosis in Europe and the United States.[14] Today, medical treatments have been developed that can suppress the virus to undetectable levels, or even substantially reduce your chance of becoming infected in the first place. These pharmaceuticals are, however, too expensive to be widely available in many parts of the world. Hence, in countries such as South Africa, HIV/AIDS is still the leading cause of mortality among all communicable, maternal, neonatal and nutritional diseases, non-communicable diseases and injuries.[15]

HIV is typically considered as a 'zoonosis'. A zoonosis (plural zoonoses) is an infectious disease caused by a pathogen or an infectious agent, such as a bacterium, virus, parasite or prion, that has jumped from an animal (typically a vertebrate) to a human host. This usually occurs when there is close contact with or consumption of the host animals. A review of infectious organisms in 2007 found 1399 species known to be pathogenic to humans, including 191 viruses and prions, 541 bacteria and rickettsia, 325 fungi, 57 protozoa and 285 helminth worms. The majority of these are zoonotic in origin.[16]

The pest most people think of as directly detrimental to human health is the mosquito. Mosquitoes transmit a wide array of pathogens, including four kinds of malaria, the dengue virus and the Zika virus. There were an estimated 219 million global cases of malaria in 2017, with a mortality rate of 28%. There are 3.4 billion people from 92 countries who are at risk of being infected. The countries most afflicted are often the poorest, including those in central Africa. Globally, we spend around US$4.3 billion each year on this disease.[17] Malaria was even, almost certainly, prevalent in England from the 15th to 19th centuries, where it was called 'the ague' or 'intermittent fever' and resulted in high levels of mortality in the fens and marshlands. The ague symptoms included anaemia; the distinctive cold, hot and sweating stages; cycling relapses; and splenomegaly or 'ague cake' (a swelling of the spleen), which could be controlled by taking quinine. The drainage of marshlands and an increasing abundance of cattle are thought to have contributed to the decline of malaria in England.[18] The increase in livestock is thought to have diverted mosquitoes away from humans, while improved housing and swamp drainage reduced the habitat for the pathogen's mosquito host.

These figures and rankings of the causes of human mortality will vary from year to year and by location. Some diseases, such as lymphatic filariasis or elephantiasis, are limited by the distribution of their insect vector. Filariasis is caused by a nematode parasite (*Onchocerca volvulus*) which is transmitted by blackflies of the genus *Simulium*. The WHO unsuccessfully listed lymphatic filariasis as a target for global elimination as a public health problem by 2020.[19]

As I write this chapter in 2021, the severe acute respiratory syndrome coronavirus 2 (SARS-CoV-2) is causing a global pandemic of coronavirus disease 2019 (COVID-19). This is almost certainly a zoonosis, probably having arisen in bats.[20] When it will end, how many lives it will take and where it will rank among the causes of human mortality remains to be seen. With a growing world population causing an increasing demand for resources and interaction with biodiversity, we should expect to see more of these 'novel' pathogens and pandemics in years to come.

## Indirect detrimental effects via crops, forestry and livestock

How do pests damage our crops? Many readers will think of insects chewing away at cabbages, or perhaps a plague of locusts in a field of corn. There are, however, an array of ways in which pests affect plants. Their impacts have been classified into different categories. There are 'stand reducers', such as pathogens that kill or weaken seeds and seedlings before or after they germinate. Growers refer to the effects of these stand reducers as 'dampening off', which is most commonly seen in wet and cool conditions. Other pests are called 'photosynthetic rate reducers', which include fungi, bacteria and viruses. 'Leaf senescence accelerators' are pathogens that result in an early decline and death of leaves and plants. Weeds can be 'light stealers' or might compete for nutrients. Finally, the insect pests appear. Some insects are called 'assimilate sappers', which are often sap-sucking arthropods such as aphids, but they can also be pathogens and nematode worms. Locusts and caterpillars are included in the final category, 'tissue consumers', which also includes vertebrates such as rabbits.

Pathogens typically play the most significant role in crop loss. One recent study provided estimates of yield losses due to pathogens and pests for five major global crops.[21] These crops were wheat, rice, maize, potato and soybean, which together are estimated to provide about half of the world's human calorie intake.

Yield loss estimates at a global level were found to be highest in rice (30.0%), followed by maize (22.5%), wheat (21.5%), soybean (21.4%) and potato (17.2%). Microscopic pathogens take first place as the leading pests causing losses in all five crops. The fungal disease sheath blight is the leading cause of crop loss in rice. Leaf rust is first among pests in wheat and is especially problematic in North and South America. It can also be devastating for grain crops in India.

In potatoes, late blight is caused by the water mould or oomycete *Phytophthora infestans*, an adaptable pathogen that is extraordinarily virulent.[22] Responsible for the Great Irish Famine, late blight has been cited as one of the most devastating

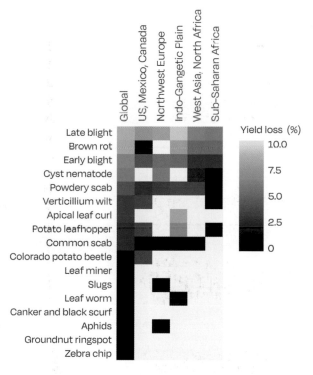

Fig. 1.2: Estimated crop losses per pathogen or pest for potatoes. The heat maps show the percentage yield losses per crop for each pathogen or pest. Losses are ranked by global totals, but the importance of different pests or pathogens varies between regions. The Colorado potato beetle has been considered for use in war; Germany and the Allies contemplated dropping them on each other's potato fields during World War II. The German government later claimed that American planes did actually drop 'Yankee beetles' over their potato fields. The German response was to set up the Potato Beetle Defence Service.

plant diseases of all time, and still costs the world as much as $6.7 billion annually in potato losses.[23] This disease in the humble Irish potato changed the world. Not only was there high mortality in Ireland, but because of this plant disease, people – including my own relatives – fled Europe and emigrated to far-flung countries such as New Zealand. Prior to the arrival of the blight disease, probably by ship from South or North America, the Irish were heavily reliant on potatoes. Supplemented by a little milk or fish, the tubers were a prominent food in the diet of approximately 40% of the Irish people. An average male labourer might consume 5 or 6 kilograms of potatoes each day. By 1851, mass death and mass emigration reduced the Irish population from 8.2 million in 1841 to fewer than

The origin of a predator can make a big difference on prey populations. Invasive species often exert a much greater level of prey suppression than native species do. For example, dingoes (*Canis dingo*; right) have co-evolved with prey species such as rock wallabies in Australia. Red foxes (*Vulpes vulpes*; left), which were introduced into Australia, have a substantially higher impact on these wallabies. It is thought that such naiveté frequently enhances suppression by alien predators. *Left: Jean-Paul Ferrero, Auscape International Pty Ltd / Alamy. Right: INTERFOTO / Alamy*

6.6 million. Most people who died probably didn't die from starvation; instead 'famine fever', which was likely to be a combination of dysentery and diarrhoeal diseases, claimed many. Death rates from 'consumption' or tuberculosis, smallpox, cholera and even measles peaked during the famine. Starvation undoubtedly contributed to the prevalence and spread of these diseases.[24]

Famine and disease go hand in hand around the globe. A slightly more recent outbreak of late blight that caused mass mortality occurred during World War I. By this time, copper had been developed into a pesticide that was sprayed to control late blight. But during the war nearly all the copper in Germany was diverted for bullets, shell casings or electric wire. In the absence of copper-based pesticides, late blight became rampant, resulting in famine for Germany and an estimated 700,000 deaths. But the humble spud wasn't absent everywhere in Europe. In Belgium, American soldiers discovered potatoes cooked in an unusual way and sold on the street. The Belgian street vendors spoke French, and French fries went on to spread around the world.[25] Late blight is still an important plant pathogen for potatoes (Fig. 1.2).

Different people have different views of 'pest' species. People who enjoy fishing in New Zealand view introduced trout as a prize species. The New Zealand Federation of Freshwater Anglers believes that the recreational trout fishery conservatively earns at least one billion dollars annually for the national economy.[26] In contrast, conservationists perceive these introduced predators as the equivalent of rats or stoats of our freshwater ecosystems, gobbling up our native fish and freshwater invertebrates alike. Colin Townsend from the University of Otago described the effects of trout at different levels, from individual to ecosystems:

> At the individual level, grazing invertebrates showed changes in behaviour as a result of the introduction of brown trout (*Salmo trutta*), a predator that exerts a very different selection pressure than do native fish. At the population level, trout have replaced nonmigratory [native] galaxiid fish in some streams but not others, and have affected the distributions of crayfish and other large invertebrates. At the community level, trout have suppressed grazing pressure from invertebrates and are thus responsible for enhancing algal biomass and changing algal species composition. Finally, at the ecosystem level, essentially all annual production of invertebrates is consumed by trout (but not by galaxiids), and algal primary productivity is six times higher in a trout stream.[27]

So, trout can clearly be considered a pest. Similarly, deer and pigs are seen by some as food and by others as invasive pests in dire need of control. Rabbits are pests to many New Zealand farmers, but pets to others (see 'Case study 1: Rabbit plagues', page 32).

More unanimous views are held for other species. The Asian tiger mosquito (*Aedes albopictus*) infects hundreds of millions of people each year with many viral pathogens, including dengue fever, the yellow fever virus and chikungunya fever, as well as several filarial nematode parasites. I'm yet to meet anyone who wouldn't describe these mosquitoes as pests and who wouldn't hesitate in slapping one to oblivion when she (only female mosquitoes drink blood) lands on their skin, seeking a meal.

An exclusion plot designed to protect vegetation from browsing by introduced populations of wallaby, deer and pigs, near Lake Ōkataina, Bay of Plenty. The browsing by these introduced species is severely depleting the forest understorey. Pigs and deer are considered an important food source for local people. *Tom Lynch photos*

## Invasive species: A special class of pests

In the 1960s game wardens on the small Pacific island of Guam noticed populations of game birds were dwindling. In the 1970s populations crashed further, without clues or carcasses to explain why. In the 1980s conservation biologists began an intensive campaign to identify the cause of wildlife declines. After considerable work, biologist Julie Savidge named her prime suspect. She had evidence that the brown tree snake (*Boiga irregularis*) was in high abundance. This nocturnal predator was attacking a wide diversity of prey on an island that had evolved without snakes, and Julie surmised it was having a massive impact. Few others initially believed her.[28]

The unintentional introduction of the brown tree snake to Guam in the 1940s has resulted in the complete loss of 10 of the 12 native forest bird species

The brown tree snake (*Boiga irregularis*) was unintentionally transported from its native range in the South Pacific to Guam, probably as a stowaway in ship cargo or the landing gear of Guam-bound aircraft, in the 1940s. These snakes have caused the extinction of at least 10 species of native birds. They cause power outages, kill domestic birds and pets, and are known to enter homes at night and attack babies and children – a behaviour that is poorly understood. A major approach on Guam to control brown tree snakes is the use of dead mice as bait, each with a tablet of paracetamol (acetaminophen), which is lethal to the snakes. Thousands of mice are produced and distributed by helicopter in canisters with streamers. *Left: Clement Carbillet, Biosphoto / Alamy. Right: US Department of Agriculture Wildlife Services*

and caused a substantial decline in populations of the two remaining species, producing a 'silent forest'. The impact of the snakes' predatory behaviour extends well beyond just the extinction of birds themselves. Many tree species require birds to disperse their seeds by eating the fruit. With the decline in birds comes a 61–92% decline in seedling recruitment.[29] Arthropod communities changed substantially as well. For example, spiders became 40 times more abundant in the presence of the snakes and the absence of birds.[30] Entire communities have changed as a result of the human-aided introduction of this snake.

A key method of snake control on Guam is now the large-scale production of mice, which are killed and placed in canisters with a tablet of paracetamol (acetaminophen), which is lethal for snakes. The canisters can be shot out of helicopters to get tangled in trees and vegetation below, where snakes find and eat the mice and paracetamol. Thousands and thousands of mice are produced and killed for this work. In New Zealand, some people are concerned about the use of 1080 to kill invasive predators; I wonder how they would feel about

A Burmese python (*Python bivittatus*) floats dead in the water after it tried, unsuccessfully, to eat an American alligator in the Everglades National Park in 2008 near Homestead, Florida. The python was introduced by accident and now competes directly with the top predators in the Everglades ecosystem. Since the introduction of these pythons, surveys have shown a 99.3% decrease in the frequency of raccoon observations, and decreases of 98.9% for opossums and 87.5% for bobcats.[31] *NPS Photo / Alamy*

paracetamol and mass-produced dead mice shot out of a helicopter . . .

The brown tree snake is a textbook example of an invasive species. Biologist Dan Simberloff describes an invasion as 'when individuals of a species not native to a region arrive with human assistance and establish an ongoing population. If the population then spreads in its new home, the phenomenon is called a biological invasion and the species is termed invasive, at least in this region.'[32] Many others will extend this definition to include the exotic species having some harmful or negative effect on the recipient population, such as the brown tree snakes' effects on bird communities. The brown trout in New Zealand's rivers, discussed above, could be considered a classic example that fits this description of an invasive species.

Quite possibly the best analysis that we have of global invasive species introductions shows a dramatic and continuous jump in species introductions over the last 200 years (Fig. 1.3). At least 37% of all first introductions occurred between 1970 and 2014. Why has this increase occurred? The authors who undertook this

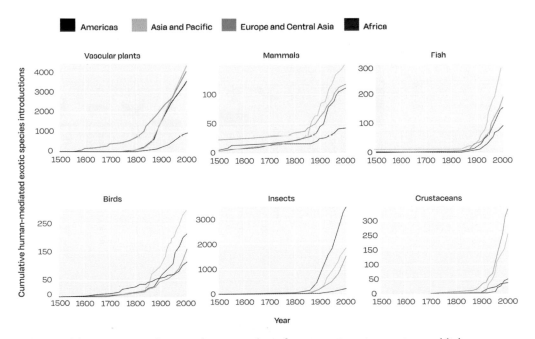

Fig. 1.3: The increase in the cumulative number of exotic or invasive species establishments shows little evidence of slowing in four regions of the world. The last 200 years has seen a dramatic rise in introduction events. *Adapted from Pyšek et al. (2020)*[33]

analysis concluded that: 'Inter-continental and inter-taxonomic variation can be largely attributed to the diaspora of European settlers in the nineteenth century and to the acceleration in trade in the twentieth century . . . This highlights that past efforts to mitigate invasions have not been effective enough to keep up with increasing globalization.' They go on to say, 'Although deleterious impacts caused by alien species have been recognized widely in legislation, there is an urgent need to implement more effective prevention policies at all scales, enforcing more stringent national and regional legislations, and developing more powerful international agreements.'[34]

Many introduced species probably have little or negligible effects. But a subset

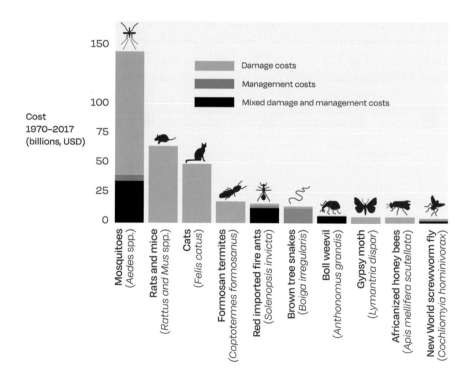

Fig. 1.4: The global economic cost of damage or management associated with invasive species is extensive. This graph shows the estimated 10 most costly taxa in billions (2017 USD), 1970–2017. Each bar represents a species or a species complex. The category of mixed costs are where costs are unable to be disentangled. Diagne et al. (2021) estimate that the total cost of all known invasions reached at least US$1.288 trillion over this period. *Adapted from Diagne et al. (2021)*[35]

of those introduced species, including the red imported fire ant and mosquitoes that carry disease and pestilence, can become major pests. Christophe Diagne and colleagues recently estimated that the global economic cost of these pests reached a minimum of US$1.29 trillion (in 2017 US$) over the period of 1970–2017 (Fig. 1.4).[35] The cost of these invasions and their management appears to treble every 10 years. Moreover, the ecological and health impacts of invasions are at least as important, but these are often incalculable.

One economic analysis of biological invasions specific to New Zealand, from 1968 to 2020, reported a total of US$69 billion in economic damage and management costs over this period. More than 80% of these costs were associated with damage, while less that 20% of this bill was invested in management. Depressingly, the costs for New Zealand are increasing, with an average cost of $US120 million in damage each year. These costs relate primarily to damage from terrestrial plants and animals.[36]

How frequently have invasive or exotic species been implicated as drivers of recent extinctions, compared with native species? Globally, invasive species have been considered a contributing cause of 33% of animal extinctions and 25% of plant extinctions. In contrast, native species were implicated in fewer than 3% and 5% of animal and plant extinctions, respectively.[37] Land transformation, climate change and the use of chemicals such as pesticides are also important drivers of extinction. But there appears to be something disproportionate about the impact of invasive or exotic species. Of the animal introductions, exotic cats, rodents, dogs and pigs have the most widespread and pervasive impacts. The endemic faunas of islands typically appear vulnerable to these invasive predators and have suffered the most from their introductions.[38]

The origin of the predators often seems to determine their effects on prey populations. For example, foxes introduced to Australia have a substantially higher impact on rock wallabies than native dingoes do. Compared with dingoes, foxes and other introduced predators suppress remnant populations of native species to a much higher degree.[39] Invasive Burmese pythons have a devastating effect on biodiversity in the Everglades National Park in Florida. The native species in the Florida region have not co-evolved with such a predator and appear unable to adapt to its sudden presence. Authors have concluded that the ecological naiveté of native species drives the larger impact of introduced predators: an ecological naiveté facilitates enhanced suppression by alien predators relative to the native enemies with which these native species have co-evolved.[40]

| Hypothesis | Description |
|---|---|
| Enemy release | Upon entry into a new range, the invasive species loses its natural enemies (herbivores, pathogens, parasites or parasitoids, and predators) that limit its population size in its home (native) range.[31] |
| Propagule pressure | The high supply and frequency of species introductions increase the chance of successful invasion due to high genetic diversity, high numbers of offspring, continual supplementation, and higher probability of introduction to a favourable environment.[41] |
| Evolution of increased competitive ability | The absence or reduction of predators that limit population in the invasive species' home range frees up resources, enabling it to enhance its competitive ability in a new ecosystem.[42] |
| Invasional meltdown | Direct or indirect symbiotic or facilitative relationships form among invasive species, causing an 'invasion domino effect'. This can occur over a range of trophic levels, where one species makes a habitat or community more amenable for the other.[43] |
| Novel weapons | Invasive species release allopatric chemicals that inhibit and repress potential competitors in the new range. Indigenous species are not adapted to the novel chemical weapons, enhancing the invader's competitive ability and success.[44] |
| Environmental heterogeneity | Habitats with high environmental variability contain a diverse array of niches that can host a variety of species. Invasion will be successful if there is an insufficient number of indigenous species to fill the available niches. This hypothesis is thought likely to occur on islands.[45] |
| Empty niche | Due to a limited indigenous species pool, the recipient, community and ecosystem are unsaturated, so invaders can use the spare resources and occupy the unused niches (i.e. there is room for the invaders). Again, this hypothesis is thought likely to occur on islands.[46] |
| Disturbance | Disturbance events increase resource availability and reset succession, giving invasive species a high chance of success at colonisation and establishment.[47] |

Table 1.1: Eight key hypotheses about why or how invasive species establish in a new region. *Adapted from Catford et al. (2009)*[48]

Invasive species may have other advantages over their native counterparts. There are diverse hypotheses to explain why some exotic species establish and become extremely abundant and widespread in their new range (Table 1.1). The 'empty niche' hypothesis, for example, is highly relevant to island nations such as New Zealand. Here, native species evolved without mammalian predators (other than some small bats). Consequently, there was a vacant niche and ample opportunity for rats or cats to establish and become highly abundant. The 'enemy release hypothesis' predicts that in their new range, invasive species benefit from a decrease in regulation by natural enemies (such as parasites) with which they had previously co-evolved. With fewer parasites, the invasive species may experience a rapid increase in abundance and distribution.[49] This hypothesis underlies pest control programmes using classical biological control, discussed in Chapter 5, whereby predators or parasites are imported to the pest's new range.

Unfortunately, however, invasive species don't always lose their parasites or pathogen cargos. Pest range expansions or invasions have frequently brought parasites and pathogens to new, naïve hosts in new areas. Over the last few centuries, for example, rat species have 'radically and explosively expanded their geographic range as a consequence of human activities'.[50] Pathogens have spread around the world with these rats, including those that cause infectious human diseases of major importance such as plague, scrub typhus, leptospirosis and hantavirus haemorrhagic fever. The accidental human-aided introduction of African mosquitoes into America that resulted in malaria outbreaks is another example of pest and pathogen co-introduction.

Clearly, not all 'exotic', 'alien' or 'introduced' species fit the description of 'pest'. One argument is that many introduced species presently fill the roles of species that humans have driven extinct (see 'Case study 2: Pablo Escobar's hippos', p. 33).

There are those that have very positive views on biological invasions. Some biologists consider that the biodiversity losses experienced around the world are offset by species introductions or hybridisation. New Zealand, for example, has a similar number of bird species now as it did prior to human arrival. The logic is that sparrows have replaced moa and blackbirds are substitutes for huia. Rats and mice are substitutes for bats. The ecological roles of rats, mice, sparrows and blackbirds are very different to those of extinct species, but some see this as an evolutionary process that is creating a 'new wild'. Fred Pearce writes that 'alien species may be scary sometimes. But they are nature at its best, and in the twenty-first century they may be its opportunity for revival after the damage done to it by humans.'[51] Scientists who support this argument consider hybridisation, for example, as a by-product of species invasions that could result in increased biodiversity. In the British Isles, there are now 88 widespread hybrid plants that have resulted from introduced and native species.[52]

The debate as to the 'pest' status of Pablo Escobar's hippos, new hybrid species and other introductions brings into play ethics and human-derived goals in biodiversity management. How do we balance the positive and negative impacts an introduced species may bring to its new distribution, especially when in some circumstances they bring negative impacts (such as fish kills from hippo dung in waterways) and benefits (such as restoration of aquatic vegetation to conditions of pre-human settlement)? What should our targets be in conservation and the management of pests and invasive species? Are there agreed-upon targets?

I started this chapter by defining a pest as any animal or plant species that is detrimental to humans or human concerns, particularly creatures that substantially damage crops, forestry or livestock, or that are simply a nuisance to people. Invasive species like mosquitoes that spread malaria, possums that kill birds in New Zealand, and brown tree snakes on Guam all fall into that category. Pathogens are clearly pestiferous and cause pestilence too, such as the bubonic plague caused by the plague bacterium *Yersinia pestis*, associated with fleas on mammalian hosts.

This reproduction from an 1882 wood engraving shows a kea (*Nestor notabilis*) attacking a sheep. In the late 1800s there were reports of 15,000–20,000 sheep being killed annually by kea. *Lyttelton Times Co Ltd. Potts, Thomas Henry: Out in the open. Christchurch, 1882. Ref: PUBL-0223-184. Alexander Turnbull Library, Wellington, NZ. /records/22843486*

For many species, the distinction isn't so black and white: they are pests to some and treasure to others. Native species can fulfil that pest definition too, such as the kākā, considered taonga (treasures) by some and pests by others (see 'Case study 3: Good parrots gone bad?', page 35). The kākā's cousin, the kea, has a similar history of being labelled a villain with a bounty on its beak. In the 1870s, reports emerged of sheep being attacked and killed by kea on high country stations in the South Island. Kea would land on a sheep's back and peck through the flesh to the kidney fat, causing wounds that were often lethal. It was estimated that between 15,000 and 20,000 sheep were dying annually. The birds were acting 'contrary to the laws prescribed to them by nature' and the consensus was for their extermination, with an estimated 150,000 killed before the bounty ended in 1970.[53] In 2021 the Department of Conservation Te Papa Atawhai estimated the total kea population to be between just 3000 and 7000 individuals.

Decisions over the labelling of kākā or kea as pests are often socio-economic issues. There is even debate over whether common bacteria that we humans host, like *Helicobacter pylori*, are a pathogen or perhaps are generally beneficial for our health. There is not likely to be a full consensus in any population. Different communities, families and individuals vary in their opinions about how detrimental or damaging a particular species might be. People's experience and understanding of a species, as well as factors such as religious or ethical beliefs, shape their attitudes towards biodiversity management.

There is no doubt that humans will continue to introduce species to new ranges. Many native species will also continue to be detrimental to humans or human concerns. And sometimes hybridisation events that produce 'new' species in the Anthropocene will produce pests or species that are perceived as beneficial.

## Pests and pestilence

In the following chapters I'll discuss ways in which we keep pests out of countries and attempt to inhibit their international movement. We will look at the different stages of a biological invasion, from native to new range, and the potential management options we can employ at each stage (Fig. 1.5). The best way to control invasive pests is to keep them outside your borders in the first place. Chapter 2 discusses the methodology and legislation designed to limit or stop the

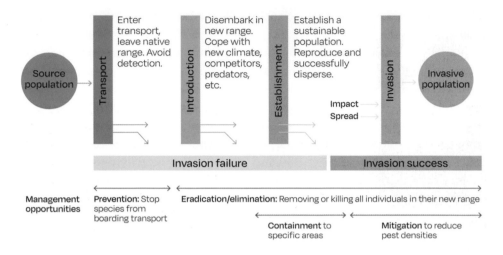

Fig. 1.5: A framework for biological invasions. Individuals of an exotic species enter a transport pathway such as a shipping container. Many will die, perhaps due to starvation or dehydration, but some will survive. If they avoid detection, they will have to cope with the new climate and species already present, reproduce and form a self-sustaining population. Many exotic species that establish in a new range have little biodiversity or economic impact, but a small proportion become pests. *Adapted from Szabo et al. (2020)[54] and Robertson et al. (2020)[55]*

unintentional introduction of species across international borders. This includes international trade and phytosanitary agreements, and targets for biodiversity conservation. Chapter 3 details New Zealand's specific biosecurity legislation, using as many examples as possible to make that less boring than it sounds.

If an exotic species does establish in a new range, eradication is a commonly attempted option, especially for species known to be highly damaging pests. Chapter 4 examines situations and conditions when biodiversity managers could or should choose eradication. Perhaps eradication is considered unachievable, so population control or limiting pest abundance might be a better option. Chapter 5 examines different methods of population control, from pesticides to gene editing, as well as public involvement and 'social licence' for different methods of pest control.

Chapter 6 gives special consideration to the management of pathogens and disease, focusing on the control of pestilence, epidemics and pandemics. With the emergence of the SARS-CoV-2 virus in 2019, which caused a global pandemic

A western chimpanzee (*Pan troglodytes verus*) suffering from leprosy. Humans are the main host for the bacterium *Mycobacterium leprae*, which causes this disease. Many animals now suffer from leprosy, including armadillos in the Americas and red squirrels in the UK. This image is from a camera trap in Côte d'Ivoire, West Africa. We don't yet know exactly how and when these chimps became infected with leprosy, but it may have been through an intermediate nonhuman disease reservoir. Humans have also spread diseases such as COVID-19 to many animal populations.

*Elena Bersacola, Cantanhez Chimpanzee Project*

of COVID-19 that has resulted in millions of deaths, New Zealand quickly adopted a 'go hard and go early' programme.[56] This included strict isolation and social distancing, which eliminated the virus from the population, saving many lives and enabling time for vaccines to be developed and delivered. Other countries have suffered much more, with substantial death tolls and economic impacts. Clearly, government involvement and science-based management and policy can make a big difference to disease management, both for biodiversity and human health. Pathogens and their effects in humans are often inextricably linked to those same diseases in biodiversity. SARS-CoV-2 is highly likely to be a zoonosis that arose in bat populations before being transmitted to and infecting humans. These zoonoses are likely to become even more frequent in our highly populated and resource-demanding world. We have, unfortunately, returned the favour, giving animals diseases that humans carry. This transferral of pathogens from humans to animals is referred to as 'reverse zoonoses'. Examples include an ongoing leprosy epidemic (caused by the bacterium *Mycobacterium leprae*) in chimpanzee populations of West Africa.[57] Human movement of diseases and pests has affected every corner of the world, even Antarctica.

In the final chapter I'll attempt to gaze into the future. Should we expect more pests? Will we see more pathogens, and animal and plant species that are detrimental to humans or human concerns? Should we expect to see more damage to our health, crops, forestry and livestock, or pests that are simply a nuisance to people? Perhaps we will see a resurgence of old pests and pathogens from our history.

Spoiler: As the world's population grows, with increasing demands and with people increasingly interacting with wildlife, the answer will certainly and unfortunately be an emphatic yes.

## Case study 1: Rabbit plagues

The first record of rabbits in New Zealand is from 1838. They were introduced by ship from New South Wales in Australia. Decades later Mark Twain would write, 'The man who introduced the rabbit into New Zealand was banqueted and lauded but they would hang him now, if they could get him.'[58]

The whole of the province of Southland and much of Otago were 'swarming' with rabbits by the 1870s. There were reports of bankrupted farmers being unable to sustain sheep, and farmers abandoning their farms. Rabbits created deserts where lush tussock had previously flourished. Family cats were kidnapped from

Rabbits (*Oryctolagus cuniculus*) around a waterhole in Wardang Island, South Australia, 1938. *National Archives of Austalia / Wikimedia Commons*

cities and sold to farmers. The profession of 'rabbiter' began, with massive harvests using poisons such as strychnine or gases that included calcium cyanide. Some farm station owners paid their staff for tokens as evidence of rabbit death. Rabbit ears were one such token, which seemed to encourage rabbiters to catch rabbits, cut off their ears and release them alive. Earless rabbits became a common sight.

In 1876 the 'Rabbit Nuisance Act' was passed into legislation. The government built an entirely ineffective rabbit-proof fence. The lack of effective control methods was lamented and prizes were offered for novel rabbit-killing devices. One such entry was a sharp spike that could be strapped to the belly of a buck rabbit, so that female rabbits would be killed while mating. Thomas Donne, a civil servant and hunter, said that a rabbit's 'laxity of morality' would be 'punished by a death wound dealt to them by the exotic weapon affixed to the buck'.[58]

Stronger action was called for and implemented through the Rabbit Destruction Council formed in 1947. This council ended the profession of rabbiter, due to the unwanted, surreptitious behaviour of rabbit farming. Selling rabbit meat became illegal. The goal of 'catching the last rabbit' was established in order to achieve the aim of completely eradicating rabbits from New Zealand. New methods were trialled, including the use of myxomatosis, and 1080, still used today. Unfortunately, to this day, New Zealand has an abundance of rabbits.

How much of an effect do rabbits have on farming? One estimate suggests that 16.5 rabbits eat as much as one stock unit (typically one ewe). High rabbit numbers can drive stocking rates (the number of animals that can graze on an area of land) from 6–13 to 0–5 per hectare.[59] Rabbits are still recognised as 'the #1 pest in Otago' by the regional council.

I grew up in Otago with a rifle in my hands and hunted rabbits regularly. An early memory from my childhood is of my mother arriving home distraught after learning of the suicide of a farmer we knew. He had left a note specifically citing rabbits as a source of a constant anguish, over many, many years, which eventually became too much.

## Case study 2: Pablo Escobar's hippos

Known as the King of Cocaine, Pablo Escobar was a Colombian drug lord and probably the richest criminal in history, due to his cartel monopolising the US cocaine trade in the 1980s and 90s. It has been estimated that 70–80 tonnes of cocaine were shipped to the United States each month by plane or submarine. His organisation would spend US$1000 each week on rubber bands to secure their cash. Rats supposedly ate 10% of these stored bills each year from warehouses.

Escobar imported one male and three female hippos from America in 1981 for his private zoo. His subsequent death and the collapse of his drug empire saw many of his animals redistributed around Colombia. But the hippos escaped both redistribution and the remnants of the zoo enclosure. Since then, the animals have bred and expanded their distribution, and their numbers now exceed 100 individuals. The conditions are near idyllic for the hippos: there are no droughts that limit population sizes as in their native Africa. They are heavy grazers of vegetation and distributors of nutrients via their faeces, and some scientists are concerned that this will alter the chemical composition of waterways and result in fish kills. The hippos also appear to alter water flow and change the landscape, and there's concern that they are displacing native species such as the manatee. As well, the hippos can be dangerous to people: they defend their territories aggressively. If we consider an invasive species to be one that is spreading and negatively affecting the biota, the 'cocaine hippos' neatly fit the definition.

There is, however, an alternative perspective. Colombia and South America once boasted many large herbivores, including a semiaquatic rhino-like creature known as a notoungulate. Giant llama roamed the landscape and their grazing and nutrient recycling probably had a major influence on the environment. Notoungulates and giant llamas have long since disappeared, largely due to us

Two hippos (*Hippopotamus amphibious*) tussle while a juvenile watches.
*Ingrid Boellaard / Alamy*

Toxodon (*Toxodon platensis*) were large, hoofed mammals that lived in South America 2.6 million to 16,500 years ago. Humans probably played a major role in their extinction. *Jaime Chirinos / Science Photo Library*

humans. Perhaps hippos are now playing the role of these extinct animals and returning the environment to a more normal or natural state?

Erick Lundgren et al. (2020) reviewed the introduction of these hippos and similar introductions around the world. According to the study: 'A primary outcome of introductions has been the reintroduction of key ecological functions, making herbivore assemblages with non-native species more similar to pre-extinction ones than native-only assemblages are.' They argue that their findings support calls for renewed research 'on introduced herbivore ecologies in light of paleo-ecological change' and that 'shifting focus from eradication to landscape and predator protection may have broader biodiversity benefits'. These authors suggest that introduced animals such as hippos are restoring the Colombian environment to what it was in pre-human times.[60]

## Case study 3: Good parrots gone bad?

It's not only exotic species that people consider to be pests.

Māori regarded kākā as rangatira (of rank or esteemed, such as a chief) among forest birds. Throughout New Zealand, flocks of hundreds of individuals could be heard (they are very loud) and seen. Kākā were hugely abundant, and an important food for Māori, and later for early European colonists. An expert hunter could catch 300 in a single day. Kākā populations were decimated by the introduction of mammalian predators such as stoats, rats and possums. In 1907, the government initiated their protection.

The establishment of the wildlife sanctuary Zealandia Te Māra a Tāne (formerly the Karori Wildlife Sanctuary) saw the reintroduction of kākā to Wellington. Fourteen kākā bred in captivity were transferred from zoos to this sanctuary between 2002 and 2007. Populations quickly grew, and individuals expanded their foraging well outside the sanctuary boundaries. I and many other Wellington residents now hear and see kākā every day. They forage, fly and call to each other even throughout the night on moonlit evenings.

The kākā (*Nestor meridionalis*) is a loud, gregarious and inquisitive native parrot. It can kill trees in its quest for food, but is also happy with a free handout.
*Piter Lenk*

An abundant kākā population, however, is not without issues. The birds forage for moth or beetle larvae in trees, and will peel bark away, sometimes entirely ringbarking and killing trees in their search for tasty insects. We have a tree on our small property that has been killed by kākā foraging. As their populations grow, kākā will undoubtedly destroy more trees. Conservation biologist Wayne Linklater has argued, 'Wellington is now a city, not a forest. Just because kākā lived here once, it does not follow logically that they should live here again. Conservationists should consider people before native species are restored.' This logic continued: 'Perhaps reintroducing kākā to a city wasn't such a good idea – a tremendous mistake by conservationists?'[61] Other conservation biologists reacted swiftly with a consensus of 'nonsense'. One suggested that 80% of Wellingtonians love kākā, despite the birds' effects on trees.[62] Many in the remaining 20% won't have an opinion on kākā, but I imagine a proportion of the city's residents will always classify these fantastic birds as 'pests'.

Some of the previously discussed species are also native plants or animals that are now commonly referred to as pests. The Colorado potato beetle used to quietly feed on native plants in North America before potatoes were introduced. Some of the major pasture pests in New Zealand are similarly native species, such as the Porina moth, *Wiseana cervinata*. Even common pathogens can vary in their pest or pestilence classification. For example, upwards of 50% of people carry the bacterium *Helicobacter pylori*, which is associated with human gastrointestinal disease, including cancers, ulcers and inflammatory disorders. But it can also confer benefits, including protection against diarrheal diseases. Exactly why it is a pathogen for some and a benefit to others is unknown.[63]

## Further reading and discussion

1. *Phytophthora infestans* is the microorganism that causes the serious potato and tomato disease known as late blight or potato blight. This plant pathogen was responsible for the Great Famine (also known as the Irish Potato Famine and the Great Starvation), a period of mass starvation and disease in Ireland from 1845 to 1852. Scientists are concerned about a resurgence of this disease. Why? What could be its effects and what can be done about it? Consider not just the effects on crops but also on the international economy. How do we ensure the future of the potato as a crop and food source?

*See:* Fry, W. E. & Goodwin, S. B. (1997). Resurgence of the Irish potato famine fungus. *BioScience, 47*(6), 363–371. doi.org/10.2307/1313151

2. The gram-negative bacterium *Helicobacter pylori* infects approximately 50% of the global human population. It is associated with a wide variety of human gastrointestinal diseases, including cancers and inflammatory disorders. Sometimes, however, it is beneficial and mutualistic. How then should we look to manage this disease and other pests or pathogens that are sometimes beneficial and sometimes harmful?

*See:* Lin, D. & Koskella, B. (2015). Friend and foe: Factors influencing the movement of the bacterium *Helicobacter pylori* along the parasitism–mutualism continuum. *Evolutionary Applications, 8*(1), 9–22. doi.org/10.1111/eva.12231

3. There is debate among scientists about the role of invasive species and even the value of the 'invasive species' label. Is this 'science denialism'? What is the road forward for the management of 'exotic', 'alien' or 'invasive' species?

*See:* Russell, J. C. & Blackburn, T. M. (2017). The rise of invasive species denialism. *Trends in Ecology & Evolution, 32*(1), 3–6. doi.org/10.1016/j.tree.2016.10.012

Davis, M.A., et al. (2011). Don't judge species on their origins. *Nature, 474*(7350), 153–154. doi.org/10.1038/474153a

4. Feral pigs in Hawai'i are considered a pest by some and a valuable food source by others. Are they really a pest? What problems do they cause? How can we mitigate the problems and satisfy those who like to hunt these animals for food?

*See:* Pejchar, L., et al. (2020). Hawaii as a microcosm: Advancing the science and practice of managing introduced and invasive species. *BioScience, 70*(2), 184–193. doi.org/10.1093/biosci/biz154

Wehr, N. H., Hess, S. C., & Litton, C. M. (2018). Biology and impacts of Pacific Islands invasive species. 14. *Sus scrofa*, the feral pig (Artiodactyla: Suidae). *Pacific Science, 72*(2), 177–198. doi.org/10.2984/72.2.1

# References

1. Becker, J. (1998). *Hungry ghosts: Mao's secret famine.* Holt Paperbacks.

2. Kreston, R. (2014, February 27). Paved with good intentions: Mao Tse-Tung's 'Four Pests' disaster. *Discover Magazine.* discovermagazine.com/health/paved-with-good-intentions-mao-tse-tungs-four-pests-disaster

3. Lampton, D. M. (1972). Public health and politics in China's past two decades. *Health Services Reports, 87*(10), 895–904. ncbi.nlm.nih.gov/pubmed/4568170

4. World Health Organization. (2017, November 15). *Plague in Madagascar.* www.who.int/emergencies/disease-outbreak-news/item/15-november-2017-plague-madagascar-en

5. World Health Organization. (2020, November 4). *Ebola virus disease.* who.int/csr/disease/ebola/en/

6. Paulson, T. (2013). Epidemiology: A mortal foe. *Nature, 502,* S2–S3. doi.org/10.1038/502S2a

7. Roth, G. A., Abate, D., Abate, K. H., et al. (2018). Global, regional, and national age-sex-specific mortality for 282 causes of death in 195 countries and territories, 1980–2017: A systematic analysis for the Global Burden of Disease Study 2017. *The Lancet, 392*(10159), 1736–1788. doi.org/10.1016/s0140-6736(18)32203-7

8. Lawlor, C. (2007). *Consumption and literature: The making of the romantic disease.* Palgrave Macmillan. doi.org/10.1057/9780230625747

9. Press Association. (2015, October 27). TB rates in parts of London 'worse than Iraq, Eritrea and Rwanda'. *The Guardian.* theguardian.com/society/2015/oct/27/tb-rates-in-parts-of-london-worse-than-iraq-eritrea-and-rwanda

10. Gao, F., Bailes, E., Robertson, D. L. et al. (1999). Origin of HIV-1 in the chimpanzee *Pan troglodytes troglodytes. Nature, 397,* 436–441. doi.org/10.1038/17130

11. Sharp, P. M. & Hahn, B. H. (2011). Origins of HIV and the AIDS pandemic. *Cold Spring Harbor Perspectives in Medicine, 1*(1), a006841. doi.org/10.1101/cshperspect.a006841

12. Faria, N. R., Rambaut, A., Suchard, M. A. et al. (2014). The early spread and epidemic ignition of HIV-1 in human populations. *Science, 346*(6205), 56–61. doi.org/10.1126/science.1256739

13. Garrett, L. (1996). *The coming plague: Newly emerging diseases in a world out of balance.* Penguin Books.

14. Huang, L., Cattamanchi, A., Davis, J. L. et al. (2011). HIV-associated *Pneumocystis pneumonia. Proceedings of the American Thoracic Society, 8*(3), 294–300. doi.org/10.1513/

pats.201009-062WR

15. Institute for Health Metrics and Evaluation. (2020, November 11). *South Africa.* healthdata.org/south-africa

16. Woolhouse, M., & Gaunt, E. (2007). Ecological origins of novel human pathogens. *Critical Reviews in Microbiology, 33*(4), 231–242. doi.org/10.1080/10408410701647560

17. Haakenstad, A., Harle, A. C., Tsakalos, G. et al. (2019). Tracking spending on malaria by source in 106 countries, 2000–16: An economic modelling study. *The Lancet Infectious Diseases, 19*(7), 703–716. doi.org/10.1016/s1473-3099(19)30165-3

18. Kuhn, K. G., Campbell-Lendrum, D. H., Armstrong, B., & Davies, C. L. (2003). Malaria in Britain: Past, present, and future. *Proceedings of the National Academy of Sciences, 100*(17), 9997–10001. doi.org/10.1073/pnas.1233687100

19. Hedtke, S. M., Kuesel, A. C., Crawford, K. E. et al. (2020). Genomic epidemiology in filarial nematodes: Transforming the basis for elimination program decisions. *Frontiers in Genetics, 10*, 1282. doi.org/10.3389/fgene.2019.01282

20. Letko, M., Seifert, S. N., Olival, K. J. et al. (2020). Bat-borne virus diversity, spillover and emergence. *Nature Reviews Microbiology, 18*, 461–471. doi.org/10.1038/s41579-020-0394-z

21. Savary, S., Willocquet, L., Pethybridge, S. J. et al. (2019). The global burden of pathogens and pests on major food crops. *Nature Ecology & Evolution, 3*, 430–439. doi.org/10.1038/s41559-018-0793-y

22. Fry, W. (2008). *Phytophthora infestans*: The plant (and R gene) destroyer. *Molecular Plant Pathology, 9*(3), 385–402. doi.org/10.1111/j.1364-3703.2007.00465.x

23. Nowicki, M., Foolad, M. R., Nowakowska, M., & Kozik, E. U. (2012). Potato and tomato late blight caused by *Phytophthora infestans*: An overview of pathology and resistance breeding. *Plant Disease, 96*(1), 4–17. doi.org/10.1094/PDIS-05-11-0458

24. Donnelly, J. S. (2002). *The great Irish potato famine.* Sutton Publishing.

25. De Jong, H. (2016). Impact of the potato on society. *American Journal of Potato Research, 93*(5), 415–429. doi.org/10.1007/s12230-016-9529-1

26. NZ Federation of Freshwater Anglers. (2020, September 7). Sports trout fishery worth 'over a billion dollars' annually threatened by trout farming proposal. *Scoop.* scoop.co.nz/stories/BU2009/S00132/sports-trout-fishery-worth-over-a-billion-dollars-annually-threatened-by-trout-farming-proposal.htm

27. Townsend, C. R. (2003). Individual, population, community, and ecosystem consequences of a fish invader in New Zealand streams. *Conservation Biology, 17*(1), 38–47. doi.org/10.1046/j.1523-1739.2003.02017.x

28. Jaffe, M. (1994). *And no birds sing: The story of an ecological disaster in a tropical paradise.* Simon & Schuster.

29. Rogers, H. S., Buhle, E. R., HilleRisLambers, J. et al. (2017). Effects of an invasive

predator cascade to plants via mutualism disruption. *Nature Communications, 8,* 14557. doi.org/10.1038/ncomms14557

30. Rogers, H., HilleRisLambers, J., Miller, R., & Tewksbury, J. J. (2012). 'Natural experiment' demonstrates top-down control of spiders by birds on a landscape level. *PLoS One, 7*(9), e43446. doi.org/10.1371/journal.pone.0043446

31. Keane, R. M. & Crawley, M. J. (2002). Exotic plant invasions and the enemy release hypothesis. *Trends in Ecology & Evolution, 17*(4), 164–170. doi.org/10.1016/s0169-5347(02)02499-0

32  Simberloff, D. (2013). *Invasive species: What everyone needs to know.* Oxford University Press.

33. Pyšek, P., Hulme, P. E., Simberloff, D. et al. (2020). Scientists' warning on invasive alien species. *Biological Reviews, 95*(6), 1511–1534. doi.org/10.1111/brv.12627

34. Seebens, H., Blackburn, T. M., Dyer, E. E. et al. (2017). No saturation in the accumulation of alien species worldwide. *Nature Communications, 8,* 14435. doi.org/10.1038/ncomms14435

35. Diagne, C., Leroy, B., Vaissière, A.-C., Gozlan, R. E. et al. (2021). High and rising economic costs of biological invasions worldwide. *Nature, 592,* 571–576. doi.org/10.1038/s41586-021-03405-6

36. Bodey, T.W., Carter, Z.T., Haubrock, P.J., et al. (2022, January 12). Building a synthesis of economic costs of biological invasions in New Zealand. *Research Square.* doi.org/10.21203/rs.3.rs-1244386/v1

37. Blackburn, T. M., Bellard, C. & Ricciardi, A. (2019). Alien versus native species as drivers of recent extinctions. *Frontiers in Ecology and the Environment, 17*(4), 203–207. doi.org/10.1002/fee.2020

38. Doherty, T. S., Glen, A. S., Nimmo, D. G. et al. (2016). Invasive predators and global biodiversity loss. *Proceedings of the National Academy of Sciences, 113*(40), 11261–11265. doi.org/10.1073/pnas.1602480113

39. Salo, P., Korpimaki, E., Banks, P. B. et al. (2007). Alien predators are more dangerous than native predators to prey populations. *Proceedings of the Royal Society B: Biological Sciences, 274*(1615), 1237–1243. doi.org/10.1098/rspb.2006.0444

40. Paolucci, E. M., MacIsaac, H. J., & Ricciardi, A. (2013). Origin matters: Alien consumers inflict greater damage on prey populations than do native consumers. *Diversity and Distributions, 19*(8), 988–995. doi.org/10.1111/ddi.12073

41. Lonsdale, W. M. (1999). Global patterns of plant invasions and the concept of invasibility. *Ecology, 80*(5), 1522–1536. doi.org/Doi 10.2307/176544

42. Blossey, B., & Notzold, R. (1995). Evolution of increased competitive ability in invasive nonindigenous plants - a hypothesis. *Journal of Ecology, 83*(5), 887–889. doi.org/Doi 10.2307/2261425

43. Simberloff, D., & Von Holle, B. (1999). Positive interactions of nonindigenous species: invasional meltdown? *Biological Invasions, 1*, 21–32.

44. Callaway, R. M., & Ridenour, W. M. (2004). Novel weapons: invasive success and the evolution of increased competitive ability. *Frontiers in Ecology and the Environment, 2*(8), 436–443. doi.org/Doi 10.1890/1540-9295(2004)002[0436:Nwisat]2.0.Co;2

45. Melbourne, B. A., Cornell, H. V., Davies, K. F., et al. (2007). Invasion in a heterogeneous world: Resistance, coexistence or hostile takeover? *Ecology Letters, 10*(1), 77–94. doi.org/10.1111/j.1461-0248.2006.00987.x

46. Hierro, J. L., Maron, J. L., & Callaway, R. M. (2005). A biogeographical approach to plant invasions: The importance of studying exotics in their introduced and native range. *Journal of Ecology, 93*(1), 5–15. doi.org/10.1111/j.0022-0477.2004.00953.x

47. Sher, A. A., & Hyatt, L. A. (1999). The disturbed resource-flux invasion matrix: a new framework for patterns of plant invasion. *Biological Invasions, 1*, 107–114.

48 Catford, J. A., Jansson, R., & Nilsson, C. (2009). Reducing redundancy in invasion ecology by integrating hypotheses into a single theoretical framework. *Diversity and Distributions, 15*(1), 22–40. doi.org/10.1111/j.1472-4642.2008.00521.x

49. Mack, R. N., Simberloff, D., Lonsdale, W. M. et al. (2000). Biotic invasions: Causes, epidemiology, global consequences, and control. *Ecological Applications, 10*(3), 689–710. doi.org/10.2307/2641039

50. Morand, S., Bordes, F., Chen, H.-W. et al. (2015). Global parasite and *Rattus* rodent invasions: The consequences for rodent-borne diseases. *Integrative Zoology, 10*(5), 409–423. doi.org/10.1111/1749-4877.12143

51. Pearce, F. (2015). *The new wild: Why invasive species will be nature's salvation*. Beacon Press.

52. Thomas, C.D. (2013). The Anthropocene could raise biological diversity. *Nature, 502*(7). doi.org/10.1038/502007a

53. Norman, G. (2018). *Birdstories: A history of the birds of New Zealand*. Potton & Burton.

54. Szabo, B., Damas-Moreira, I., & Whiting, M. J. (2020). Can cognitive ability give invasive species the means to succeed? A review of the evidence. *Frontiers in Ecology and Evolution, 8*, 187. doi.org/10.3389/fevo.2020.00187

55. Robertson, P. A., Mill, A., Novoa, A., et al. (2020). A proposed unified framework to describe the management of biological invasions. *Biological Invasions, 22*(9), 2633–2645. doi.org/10.1007/s10530-020-02298-2

56. Jamieson, T. (2020). 'Go hard, go early': Preliminary lessons from New Zealand's response to COVID-19. *The American Review of Public Administration, 50*(6–7), 598–605. doi.org/10.1177/0275074020941721

57. Hockings, K.J., Mubemba, B., Avanzi, C. et al. (2021). Leprosy in wild chimpanzees. *Nature, 598*, 652–656. doi.org/10.1038/s41586-021-03968-4

58. Druett, J. (1983). *Exotic intruders: The introduction of plants and animals into New Zealand*. Heinemann.

59. Latham, A. D. M., Latham, M. C., Norbury, G. L., et al. (2019). A review of the damage caused by invasive wild mammalian herbivores to primary production in New Zealand. *New Zealand Journal of Zoology, 47*(1), 20–52. doi.org/10.1080/03014223.2019.1689147

60. Lundgren, E. J., Ramp, D., Rowan, J., et al. (2020). Introduced herbivores restore Late Pleistocene ecological functions. *Proceedings of the National Academy of Sciences, 117*(14), 7871–7878. doi.org/10.1073/pnas.1915769117

61. Linklater, W. (2016, May 10). Kaka conflict: Conservation icon to pest. *Dominion Post*.

62. Daugherty, C. (2016, May 23). Too many kaka? What nonsense. *Dominion Post*.

63. Lin, D., & Koskella, B. (2015). Friend and foe: factors influencing the movement of the bacterium *Helicobacter pylori* along the parasitism-muualism continuum. *Evolutionary Applications, 8*(1), 9–22. doi.org/10.1111/eva.12231

# 2. KEEPING PESTS AND DISEASES OUT

## The best approach is not to allow them to cross your borders

In the 1950s, a large and very alive snake was shipped to New Zealand in a car from Australia. There are no snakes in New Zealand, so it was abundantly clear that this was a border incursion and an unintentional import. The car's owner discovered the snake after hearing some rustling on the back seat, and on looking behind saw the coiled serpent staring back at him. The driver didn't panic. Instead, he took the snake to the local police station, where he tossed the poor reptile inside. The police constables inside reacted in horror and leapt onto their desks, until a calmer sergeant caged the snake with an upturned wastepaper basket. He then wrote his report with his feet resting on the basket. (His relaxed demeanour was broken, unfortunately, when a constable 'crept up behind him and gently ran his hand up the sergeant's propped-up leg'.[1])

New Zealand has seen many other snake incursions. A 2021 arrival was a metre-long juvenile non-venomous carpet python (*Morelia spilota*), also shipped from Australia but found this time in a pipe at a construction site. A spokesperson for the Ministry for Primary Industries (MPI) told the media that they 'intercept one or two snakes a year'. The spokesperson went on to say the reptiles typically aren't alive: 'They are normally not venomous and mostly arrive dead, due to treatment of imported cargo.' Most snake species would probably find New Zealand too cold. 'However, a snake from a cooler area would have a chance of survival. If such a snake were carrying eggs, there is a very small chance that the offspring could survive.'[2] Snake introductions can have considerable consequences. The brown tree snake invasion in Guam and the Burmese python introduction into the Florida Everglades have caused major losses in biodiversity, as discussed in Chapter 1. These snakes are now well established in their new environments. Eradicating them, as much as people might try, seems unlikely, and any attempts to do so will be hugely expensive.

The best approach for managing exotic species and invasive species, whether they are snakes, plants, viruses, fruit flies or rats, is not to let them into your territory at all. New Zealanders call this 'biosecurity', which the government defines as the exclusion, eradication or management of pests and diseases that pose a risk to the economy, environment, and cultural and social values, including human health.[3] Other countries and organisations get a bit more technical. Biosecurity, as defined by the Food and Agriculture Organization of the United Nations (FAO), 'offers a strategic and integrated approach to analyse and manage **risks** in food safety, animal and plant life and health, and biosafety. It provides a policy and regulatory framework to improve coordination and take advantage of the synergies that exist across sectors, helping to enhance protection of human, animal and plant life and health, and facilitate **trade**.'[4] The words *risks* and *trade* in that definition are important, because managing both is critically important for limiting exotic species introduction. There is considerable risk to trade, but people want to travel and to import and export goods around the globe. I'd like a fancy, reliable, Japanese-made car and perhaps a holiday to see the wildlife in

A Burmese python in Florida. Their eradication from the US is being considered, but it would be extraordinarily difficult and expensive using current pest control techniques. Hindsight is always 20/20. The better and easier approach would have been not to let these snakes across international borders to reach Florida in the first place. *Austin Fitzgerald photo*

the Amazon rainforest. I need my coffee in the morning and like a square or two of chocolate in the evening. There are risks of unintentional organisms as passengers with those coffee, chocolate and car imports. There is a risk that while on my holiday I'll acquire and return home with one of the many diseases in the Amazon that aren't established in New Zealand, including malaria, dengue, Chagas disease or even leprosy. The degree of risk that governments and citizens are willing to tolerate is variable, as are the costs and level of treatments needed for unintentional passengers.

In this chapter I will review the international conventions and agreements that have attempted to achieve biosecurity between borders, with an emphasis on international trade, health, and the movement of goods and people (Table 2.1). There is a tightrope to walk. Finding the balance of trade and economic growth while minimising the risk of invasive species or pathogen entry is extraordinarily difficult, as any level of international trade brings some degree of invasive species risk. How do we manage and minimise this risk? What are the international procedures and regulations regarding human health? I'll also discuss how governments have invoked 'biosecurity' for nefarious means: imposing trade barriers based on dubious scientific evidence in order to protect their citizens' industries.

## Historic trade and early developments towards international agreements

Every time I am in a major port city, I watch ships coming and going, loading and offloading container after container, and wonder why the world doesn't have a completely homogeneous fauna. There is so much international movement of goods and people. How is it that species that are likely to spread haven't spread already?

Early intercontinental explorers were some of the first to see the devastating effects of their movement and trade. Contrary to popular opinion, guns and fierce soldiers played only a minor role in the European colonisation of North America. Instead, the common childhood illnesses brought from the Old World by the European conquistadors were the major tools of conquest. Diseases such as smallpox, measles and typhus annihilated most of the American Indigenous populations. Peru suffered through as many as 17 epidemics between 1520 and 1600, with horrific scenes and reports that 'the Indians die so easily that the bare look and smell of a Spaniard causes them to give up the ghost'.[5] The explorers, conquistadors and colonisers often knew they were carrying diseases and invasive species: the deliberate infection of Indigenous peoples was a form of biological warfare. One of the most famous but debated examples of intentional transmission was at Fort Pitt in Pennsylvania in 1763, involving smallpox – a horrible disease caused by two virus variants, *Variola major* and *Variola minor*. With a heavily populated fort under siege, 'two blankets and a handkerchief' from the fort's smallpox hospital were provided to the surrounding local Indians in 'the hope that it will have the desired effect'. Authors including Fenn (2000) conclude that such biological warfare, often involving smallpox, was a widespread reality in eighteenth-century North America, and was designed for 'the ruin of a whole country, involving the indiscriminate murder of women and children'.[6]

Many other explorers, perhaps with more innocent or ignorant intent, were happy to see vermin scuttle off their ships when anchored near land. Captain James Cook noted that his ship was 'a good deal pestered with rats', and rid himself of at least a few when they ventured ashore along cables and planks in Tahiti in 1785.[7] However, zoonotic diseases such as typhus follow animal hosts (such as rats) and their associated vectors (such as lice and fleas).

*HMS 'Resolution' and 'Discovery' at Moorea*, a painting by John Cleveley of ships on James Cook's voyage to Tahiti in 1778. Many diseases were introduced around the world by early explorers, including smallpox, measles, cholera and influenza. *Royal Museums Greenwich / Wikimedia Commons*

## Table 2.1: Key international agreements relevant to biosecurity*

### General Agreement on Tariffs and Trade (GATT), 1947

Promotes international trade by reducing or eliminating trade barriers such as tariffs or quotas. The World Trade Organization (WTO) is the successor to the GATT, though the original text is still in effect. Article 20 of GATT allows nations to act on trade to protect human health (and morals!), and animal and plant health, provided they do not discriminate or use the provision as disguised protectionism. One hundred and sixty-four countries are signatories. Eritria, North Korea are non-members. Several other North African countries, and Iran and Iraq, have 'observer' status.

### United Nations Convention on the Law of the Sea (UNCLOS), 1982

Defines the rights and responsibilities of nations with respect to their use of the world's oceans, establishing guidelines for businesses, the environment, and the management of marine natural resources. A key provision is to define ocean boundaries, though the convention establishes general obligations for safeguarding the marine environment and protecting the freedom of scientific research on the high seas. The convention has been ratified by 168 parties, which includes 167 states. Non-parties include the US, Peru and South Sudan.

### Convention on Biological Diversity (CBD), 1993

Its three main goals: the conservation of biodiversity, the sustainable use of its components, and the fair and equitable sharing of benefits arising from genetic resources. Two supplementary agreements are the Cartagena Protocol (governing the movements of living modified organisms resulting from modern biotechnology from one country to another) and the Nagoya Protocol (regarding the fair and equitable sharing of benefits arising from the utilisation of genetic resources). The CBD has 196 signatories, including all but one United Nations member state. The US is the one non-signatory member. Signatories are obliged, not legally bound.

### Sanitary and Phytosanitary Measures (SPS Agreement), 1995

An international treaty developed by the WTO. Its measures aim to protect human, animal and plant life from certain risks. Three organisations set standards that WTO members should base their SPS methodologies on: Codex

Alimentarius Commission (Codex), World Organization for Animal Health (OIE), and the Secretariat of the International Plant Protection Convention (IPPC). The signatories are the same as for the GATT and WTO.

## Public Health Emergency of International Concern (PHEIC) and International Health Regulations (IHR), 2007

A PHEIC is a formal declaration by the WHO of 'an extraordinary event which is determined to constitute a public health risk to other States through the international spread of disease and to potentially require a coordinated international response'. It is initiated when a 'serious, sudden, unusual, or unexpected' situation arises that 'carries implications for public health beyond the affected state's national border' and 'may require immediate international action'. The 2014 outbreak of Ebola in West Africa and the ongoing COVID-19 pandemic are examples. The IHR coordinate these responses, and are binding on 196 states parties, including all 194 member states (countries) of the WHO.

## Aichi Biodiversity Targets, 2010

The 10th meeting of the Conference of the Parties for the CBD, in Nagoya, adopted a revised Strategic Plan for Biodiversity, including the Aichi Biodiversity Targets, for 2011–2020. The goal was to take urgent action to halt the loss of biodiversity. Target 9 is especially relevant for biosecurity: 'By 2020, invasive alien species and pathways are identified and prioritized, priority species are controlled or eradicated, and measures are in place to manage pathways to prevent their introduction and establishment.' The signatories are as for the CBD.

## Convention on the International Maritime Organization, 2011

The 2011 Guidelines for the Control and Management of Ships' Biofouling to Minimize the Transfer of Invasive Aquatic Species are meant to provide a globally consistent approach to the management of biofouling, which is the accumulation of various aquatic organisms on ships' hulls. They were adopted by the Marine Environment Protection Committee (MEPC) in July 2011.

\* Some conventions with relevance to international biosecurity risk are not included, including the Convention on Wetlands, World Heritage Convention and the Convention on International Trade in Endangered Species.

We've come a long way since the early explorers and colonists, although so has the movement of people and the volume of trade that we send around the world. The United Nations World Tourism Organization (UNWTO) estimates that in the year 1950 there were just 25 million international tourists. By 2018, this number had increased to 1.4 billion people travelling internationally.[8] In regards to shipping, the International Chamber of Shipping reports that 11 billion tonnes of goods are transported by ship each year, which represents ~1.5 tonnes per person based on the current world population.[9] Seaborne trade has quadrupled in size over the last four decades. The immensity of shipping was highlighted when the container ship MV *Ever Given* became stuck in the Suez Canal in 2021. This single 400-metre-long ship can carry 20,000 containers at any time.

Air, road and rail transport are in addition to shipping transport. This massive movement of goods and people, and the potential risks for biosecurity, have seen the development of several key international health or trade agreements.

Perhaps unsurprisingly, the first set of international meetings regarding

The Shanghai Yangshan deep-water port for container ships. The scale of the modern-day movement of goods around the globe presents a major risk of invasive species introductions. *Roland Nagy / Alamy*

international quarantine regulations were in regard to human health. The so-called International Sanitary Conferences were a series of 14 conventions between 1851 and 1938 that focused on standardising international quarantine regulations against the spread of just three diseases: cholera, bubonic plague and yellow fever. The first was organised by the French Government in 1851 around the key question of whether cholera should be subject to quarantine regulations. Cholera epidemics and pandemics motivated many of the meetings. The first conference exclusively on bubonic plague was in 1897. This was the time of the third plague pandemic, after it appeared to have re-emerged from a wild rodent reservoir in the remote Chinese province of Yunnan in 1855. This plague had spread to and was rampant in Bombay (now known as Mumbai), India, by 1896. By 1900 it had reached ports on every continent, carried by infected rats travelling the international trade routes on the new steamships. The role of international transport of rats and their fleas in spreading the disease was highlighted in 1900, when the plague arrived in Sydney and spread out from the Darling Harbour wharves. The pandemic officially ended in 1959, by which time the global death toll stood at more than 15 million people.[10]

Professional rat-catchers during the outbreak of bubonic plague in Sydney, 1900. From 1900 to 1925 there were 12 major outbreaks in Australia as ships imported wave after wave of infection. An estimated 535–1371 people in Australia died from plague over this period.[11] *State Library of NSW / Wikimedia Commons*

The International Sanitary Conferences started with a European focus, with representation from just 12 countries. The meetings grew in global representation and were a major impetus for the formation of the WHO in 1948. One of the key goals and potential benefits of the establishment of the WHO was that a single set of infectious disease rules would be able to be administered by one international health organisation with near global universal membership.[12]

The period immediately after World War II was productive for agreements in both trade and health. The General Agreement on Tariffs and Trade (GATT) was a product of a United Nations Conference on Trade and Employment (Table 2.1). It was signed by a small number of countries in 1947; that membership has since grown to 164 nations. The World Trade Organization (WTO) is the successor to the GATT, though the original text is largely still in effect. A key goal of the GATT was to reduce or eliminate trade barriers such as tariffs or quotas. Biosecurity was, however, noted as a concern in the agreement, primarily to balance trade and public health. The GATT included a general trade exception that enabled countries to violate a GATT obligation if this violation was necessary to protect human life or health. The GATT attempted a dispute resolution process, although under this agreement a consensus was required on a ruling, and the losing party would not usually consent. There was no Appellate Body under the GATT.

You can, if you want, download the six-volume set of *GATT Disputes: 1948–1995*.[13] Of the 316 disputes over the period, the United States was the complainant in 93 and the respondent in 91. One of the more famous cases addressed the United States' controversy with the European Communities over hormones used in the production of beef. This case has been referred to as 'the beef war'. The synthetic version of the hormone estradiol was particularly controversial, as it is the major human female sex hormone. As context for this time and debate, a series of 'hormone scandals' had occurred in Europe relating the onset of premature puberty in school children to the consumption of hormone-treated beef. Consequently, in 1989 the European Communities implemented a ban on the import of all beef products treated with growth hormones. The ban was based on an assessment that there was no scientific evidence guaranteeing that beef treated with growth hormone was risk free, and thus Europe should have a right to adopt a precautionary ban to protect its citizens. The European nations had and still have a reputation for a stringent application of the precautionary principle to food safety.

The precautionary principle or approach was developed in Germany in the 1970s in regard to environmental protection. A broad epistemological, legal and philosophical approach to situations that have the potential to cause harm, it is often applied when scientific knowledge is lacking. It emphasises caution, taking time to review before accepting and implementing innovations, products or services that may have deleterious consequences. It came into global use at the UN Conference on Environment and Development in Rio de Janeiro in 1992. The Rio Declaration asserts: 'Nations shall use the precautionary principle to protect the environment. Where there are threats of serious or irreversible damage, scientific uncertainty shall not be used to postpone cost-effective measures to prevent environmental degradation.' Opponents argue that the precautionary principle is often self-cancelling, unscientific and an obstacle to progress.

The United States protested the ban against the import of beef products treated with growth hormones for the reason that it was not based on scientific evidence and so was a disguised barrier to trade. The case was heard at the WTO, which in 1998 found that, in the absence of a risk assessment, there was no valid case to prohibit the importation of North American beef. The WTO authorised the United States to implement retaliatory tariffs. Beef exports, tariffs and trade were subsequently contentious for decades. Risk assessments were performed and argued over, and memorandums of understanding were signed, before a duty-free tariff rate quota was agreed upon in 2019.

## Sanitary and Phytosanitary Measures (The SPS Agreement)

The GATT had a focus on lowering tariffs on international trade. The agreement was successful in this ambition. There were, however, other non-tariff barriers to trade. Under the GATT, countries could legitimately stop the trade of a specific commodity on the basis that officials were concerned over the introduction of invasive species or disease. Alternatively, and more nefariously, a country might use the presence of an invasive species in the export nation as an excuse for a trade barrier to protect its own producers from economic competition. Clearly, there needed to be rules regarding the right to protect your territory from disease and pests while also limiting the potential for unreasonable and protectionist non-tariff barriers.

The measures in the WTO Agreement on the Application of Sanitary and Phytosanitary Measures (SPS Agreement) are aimed at the protection of human, animal or plant life or health (Table 2.1). It came into force in 1995. The SPS Agreement covers not only invasive species and diseases, but also levels of precaution that could be implemented around pesticide, chemical and bacterial contamination. Countries can implement quarantine or biosecurity precautions, but these must be based on scientific principles and sufficient scientific evidence. Risk assessments are key. Signatories to the agreement must consider all scientific evidence in their risk assessments, including the prevalence of pests and diseases, and the efficacy of quarantine measures. Unfortunately, for many pests or diseases there is little information available. In these data-deficient circumstances, countries can adopt sanitary or phytosanitary measures based on available information, but are obliged to seek more information to develop a better risk analysis within a 'reasonable time period'.[14] Countries can require quarantine measures such as the fumigation of logs to prevent the spread of insect pests.

Logs at Port Marlborough, Picton, with methyl bromide fumigation going on beneath the black plastic sheets. Methyl bromide is used for the eradication of pest species from import cargo. New Zealand used 525 tonnes of it in 2014, most on forestry products. The recapture of the gas has been required since 2020. Because of its ozone-depleting nature, it is to be phased out under the Montreal Protocol for non-quarantine use. *Tim Cuff / Alamy*

There have been several high-profile cases regarding the implementation of the SPS Agreement. The previously discussed 'beef war' about the North American export of hormone-treated beef began in 1989 under the GATT and ended with negotiations under the SPS Agreement.

Another long-running example is a dispute over the export of fresh salmon from Canada to Australia. In 1975 Australia banned imports of fresh salmon from Canada based on diseases that might damage the Australian salmon-farming industry. Canada argued that this Australian ban was simply economic protectionism; that there was no evidence that fresh imports of salmon posed any significant risk. A WTO dispute resolution ruled in favour of Canada in 1997. They found that Australia's measures were inconsistent with Articles 2.2 and 2.3, which obligate countries to base decisions on scientific evidence and that ensure the implementation of sanitary and phytosanitary measures do not arbitrarily or unjustifiably discriminate. Australia's actions were also inconsistent with Articles 5.1, 5.5, and 5.6, which require appropriate risk assessments. After appeals, the case was finally concluded in favour of Canada in 2000, 25 years after the case was initiated.[15]

Australia was also the focus of a case involving New Zealand and the bacterial disease fire blight (see 'Case study 2: The WTO, the SPS Agreement and NZ exporting apples to Australia', page 73).

## The SPS Agreement and the three 'sister' standards for WTO members

The SPS Agreement has an extremely broad focus that includes invasive species and diseases, and pesticide, chemical and bacterial contamination. This diversity of issues requires a range of expertise and advice. Countries that are members of the WTO base their SPS methodologies on three 'sister' standards developed by three different organisations: the World Organisation for Animal Health (known by their historical acronym OIE), the Secretariat of the International Plant Protection Convention (IPPC), and the Codex Alimentarius Commission (Codex).

The need to combat animal diseases at a global level led to the creation of the Office International des Epizooties in 1924. In 2003 it became the World Organisation for Animal Health (OIE). It is now the intergovernmental organisation responsible for improving animal health worldwide, with the goals of ensuring transparency in the global animal disease situation; collecting,

An engraving depicts an outbreak of rinderpest in Holland in 1745. The disease struck in southern Africa in the 1890s, killing 80–90% of all cattle in eastern and southern Africa. *Jan Smit / Wikimedia Commons*

analysing and disseminating veterinary scientific information; and encouraging international solidarity in the control of animal diseases.[16] The OIE's role is to safeguard world trade by publishing health standards for international trade in animals and animal products. The OIE provides recommendations for improving the legal framework and resources of national veterinary services, to provide a better guarantee of food of animal origin and to promote animal welfare through a science-based approach. A key resource is a single OIE list of notifiable terrestrial and aquatic animal diseases.[17] In 2021 this list contained 117 diseases or pathogens. The prion disease Scrapie, discussed in the next chapter, is among these diseases (see 'Case study 1: Prions, mad cows and sheep, and cannibalism', page 111). Also included in the OIE list is the viral disease rinderpest. The devastation caused by rinderpest was the impetus for the very first medical school in France in 1761 and the creation of the OIE in 1924. The last known case of rinderpest was in 2001, and global eradication was declared in 2011. This represents an extraordinary milestone in the history of animal health and for the OIE. The only comparable achievement is the eradication of the human viral disease smallpox, which was achieved in 1980. At least, we hope smallpox is gone for good. There are fears that this virus could again rise. The last recorded fatal case of smallpox was Janet Parker, who died in Birmingham, England, in September 1978. She was infected by the virus after it 'escaped' from a laboratory. Some laboratories around the world still have stocks of the smallpox virus, causing concerns that the virus could again escape or even be used for bioterrorism.[18]

The IPPC is a treaty that aims to protect the world's plant resources from the introduction and spread of pests, and to promote safe trade. It is signed by 180 countries. The Secretariat of the IPPC develops standards and recommendations regarding pest management. You can find relatively simple standardised guidelines on how to determine the pest status in a particular area,[19] a framework for pest risk analysis (Fig. 2.1)[20] and even guidelines for pest eradication programmes.[21]

In 2021 the Secretariat released a report on the impact of climate change on plant pests (see 'Case study 3: The International Plant Protection Convention report on the impacts of climate change on plant pests', page 75). A key, depressing finding from their report was: 'The majority of studies indicate that, in general, pest risk will increase in agricultural ecosystems under climate change scenarios, particularly in today's cooler Arctic, boreal, temperate and subtropical regions. This is largely true for forestry pest risks as well.'[22]

The final standard that underpins the SPS Agreement is the Codex Alimentarius, developed by the FAO. The Codex is a collection of standards, codes of practice,

A child suffering from smallpox. The disease had a mortality rate of approximately 30% for all infected cases. The last known natural case was in 1977. Global eradication was declared in 1980. The WHO describes the global eradication of smallpox as among the most notable and profound public health successes in history.
*Centers for Disease Control and Prevention*

guidelines and other recommendations involving food, food production, food labelling and food safety. The Codex Alimentarius has an important role to 'provide effective, science-based, internationally accepted standards of food safety . . . as well as to facilitate international food and agricultural trade'.[23] It has less to do with pest management than the IPPC and the OIE. As with probably every set of recommendations ever made on trade, the application of the Codex has not been without debate. For example, there has been considerable argument on how

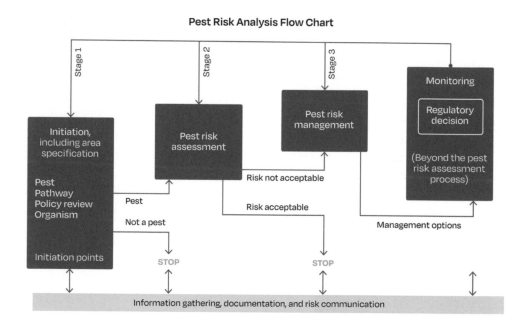

**Pest Risk Analysis Flow Chart**

Fig. 2.1: Flowchart from the IPPC's framework for pest risk analysis. Risk analysis might be initiated when importation is proposed of a commodity not previously imported or a commodity from a new area of origin. Species are considered potential pests if they are repeatedly intercepted at import or identified as potentially damaging by research or observation by stakeholders. Stage 3 involves the identification of phytosanitary measures that (alone or in combination) reduce the risk to an acceptable level. The conclusion of the pest risk management stage will depend on whether appropriate phytosanitary measures are available, cost-effective and feasible. Monitoring the outcome of the management and regulatory decision is highly recommended. Risk communication is recognised as an interactive process, allowing exchange of information between the NPPO and stakeholders.

vitamin supplements should be classified and regulated. A German delegation in 1996 proposed that no vitamin, herb or mineral should be sold for preventive or therapeutic reasons. Their proposal that supplements should be reclassified as drugs was passed, but its implementation was halted after protests. Some of the more extreme conspiracy theories regarding the Codex have suggested it to be 'a complicated plan involving GMOs, Big Pharma, Nazi war criminals, the banning of vitamins and holistic health supplements, and corporations taking control of the world's food supply – with the ultimate goal being the depopulation of the planet through toxic food and malnutrition'.[24] There are crazy people everywhere.

## Convention on Biological Diversity

People around the world argue or debate the need for pest control and whether to manage invasive species. To a large extent, however, these arguments are moot. As signatories to the 1992 Convention on Biological Diversity (CBD),[23] New Zealand is one of 196 countries around the globe that are obligated to manage their acquired invasive species. One of the three objectives of this convention is to conserve biological diversity, which obliges the signatories to prevent the introduction of, and control or eradicate, alien species that threaten ecosystems, habitats or species.

A revised and updated Strategic Plan for Biodiversity was produced in 2010 from the tenth meeting of the Conference of the Parties. This plan included the Aichi Biodiversity Targets (Table 2.2), named for the Aichi Prefecture of Japan where the conference was held. There were 20 targets, with the overall goals of halting biodiversity loss and ensuring that ecosystems continue to provide essential services. Target 12 stated: 'By 2020 the extinction of known threatened species has been prevented and their conservation status, particularly of those most in decline, has been improved and sustained.' Target 9 was: 'By 2020, invasive alien species and pathways are identified and prioritized, priority species are controlled or eradicated, and measures are in place to manage pathways to prevent their introduction and establishment.'[25] Consequently, there was a requirement for most of the world's countries to prioritise their invasive pests and manage them based on an evaluation of harm.

The Aichi Biodiversity Targets were for the period 2011–20. Were they successful? Have we controlled or eradicated invasive species and prevented

**Vision:** Living in harmony with nature where by 2050, biodiversity is valued, conserved, restored and wisely used, maintaining ecosystem services, sustaining a healthy planet and delivering benefits essential for all people.

**Mission:** Take effective and urgent action to halt the loss of biodiversity in order to ensure that by 2020 ecosystems are resilient and continue to provide essential services, thereby securing the planet's variety of life, and contributing to human well-being, and poverty eradication....

### Strategic Goal A — Address the underlying causes of biodiversity loss by mainstreaming biodiversity across government and society

- Awareness of biodiversity increased
- Biodiversity values integrated
- Incentives reformed
- Sustainable production and consumption

### Strategic Goal B — Reduce the direct pressures on biodiversity and promote sustainable use

- Habitat loss halved or reduced
- Sustainable management of aquatic living resources
- Sustainable agriculture, aquaculture and forestry
- Pollution reduced
- Invasive alien species prevented and controlled
- Ecosystems vulnerable to climate change

### Strategic Goal C — Improve the status of biodiversity by safeguarding ecosystems, species and genetic diversity

- Protected areas
- Reducing risk of extinction
- Safeguarding genetic diversity

### Strategic Goal D — Enhance the benefits to all from biodiversity and ecosystem services

- Ecosystem services
- Ecosystem restoration and resilience
- Access to and sharing benefits from genetic resources

### Strategic Goal E — Enhance the benefits to all from biodiversity and ecosystem services

- Biodiversity strategies and action plans
- Traditional knowledge
- Sharing information and knowledge
- Mobilizing resources resources from all sources

**Implementation, Monitoring, Review & Evaluation**
- The provision of financial resources
- National biodiversity strategies and action and national and regional targets
- The participation of all relevant stakeholders
- Supported and encouragement of initiatives and activities of indigenous and local communities
- The Convention's programmes of work

**Support Mechanisms**
- Capacity-building for effective national action
- Clearing-house mechanism and technology transfer
- Financial resources
- Partnerships and initiatives to enhance cooperation
- Support mechanisms for research, monitoring and assessment

Table 2.2: Strategic Plan for Biodiversity 2011–20 (the Aichi Biodiversity Targets). *Adapted from the Global Biodiversity Outlook 5*[25]

further introductions? An assessment from the Secretariat of the CBD concluded: 'The projected growth in global shipping traffic is likely to increase the risk of alien species invasions by between three and 20 times the current level by 2050. The increased risk is forecast to be especially high in middle-income countries, notably in Northeast Asia. Shipping growth is anticipated to have a far greater effect on marine invasions than climate-driven environmental change.'[25] On the positive side, there has been some progress, including advances in identifying and prioritising invasive species, and some success in eradicating populations of invasive species. We are getting better and better at removing invasive species from islands, resulting in major conservation gains.[26] These successes, however, represent only a fraction of all occurrences of invasive species. There is no evidence of a slowing down in the number of new introductions of alien species. This Aichi Biodiversity Target has only been partially achieved.

The report concludes that Aichi Biodiversity Target 12, regarding preventing extinction of threatened species, has not been achieved. 'Species continue to move, on average, closer to extinction . . . Among well-assessed taxonomic groups, nearly one-quarter (23.7%) of species are threatened with extinction unless the drivers of biodiversity loss are drastically reduced, with an estimated total of one million threatened species across all groups. Wild animal populations have fallen by more than two-thirds since 1970, and have continued to decline since 2010.'[25] There is a growing need for quick action, with eradications of invasive pests as a high priority, before we see more extinctions (Fig. 2.2).

New Zealand has responded to the Aichi Biodiversity Targets with varying degrees of success. There has been an increasing programme to manage biosecurity risks at the border by surveillance and monitoring. The eradication of some invasive species has occurred, such as rat eradications on offshore islands. In New Zealand's *Sixth National Report to the United Nations Convention on Biological Diversity* in 2019, the Predator Free 2050 programme to eradicate possums, rats and stoats in New Zealand by 2050 was emphasised as a step towards meeting goals on eradication and invasive species control. The assessment provided by the Department of Conservation Te Papa Atawhai on threatened species, however, paints a less positive picture. An assessment of ~12,000 indigenous organisms showed that only 24 'threatened and at risk taxa' have improved in their conservation status, while populations of 87 other species have declined or moved into a worse status. The decline of many of these indigenous and valued species is due to invasive species. The evaluation of New Zealand's response to the Aichi

Biodiversity Targets is that our terrestrial and aquatic ecosystems continue to face substantial and significant pressures. The restoration programmes implemented in New Zealand are considered not to have delivered significant improvements – at least not yet.[27]

Fig. 2.2: The decline in birds on Gough Island, a 91-km² World Heritage Site in the South Atlantic, is attributed solely to house mice, the only introduced mammalian predator. Mice were introduced there by sailors in the 19th century and have evolved to be much larger than normal. The annual breeding success for Tristan albatrosses (*Diomedea dabbenenais*) has been found to be as low as 27%; breeding success for other *Diomedea* species of albatross typically ranges from 60–75%. Mouse predation appears sufficient to drive extinctions on this island unless it is managed.[44] A mouse eradication programme was begun in 2021. *Kim Haddrell / Royal Society for the Protection of Birds*

## Management of biofouling: The Convention on the International Maritime Organization

Biofouling and ballast water have been identified as a major pathway for invasive species in the marine environment around the globe. The diversity of exotic species as biofouling pests can be staggering. One comprehensive analysis of a single ship intercepted in the North American Great Lakes found 944 living, biofouling individuals with at least 74 distinct marine and freshwater taxa.[28] Ballast water has been the likely source of approximately 75% of the exotic and invasive species introduced into the Great Lakes.[29]

One of the most disastrous cases of the international movement of aquatic pest species via ballast water is the zebra mussel (*Dreissena polymorpha*). This mussel is native to the lakes of southern Russia and Ukraine and was introduced via ballast water into North America in the 1980s. One study in 2007 estimated that zebra mussels have cost North American electric-generation and water-treatment facilities US$267 million over the period of 1989–2004.[30] Ecologically, zebra

The zebra mussel (*Dreissena polymorpha*) is native to the lakes of southern Russia and Ukraine. It is thought to have been introduced via ballast water into North America through the Great Lakes. Here, dozens of zebra mussels have settled on a painter's mussel (*Unio pictorum*). Zebra mussels smother and kill many native clams and mussels in North America and are driving some species to near extinction. They block pipelines and water intakes, and are believed to be the source of avian botulism. On the right, an oil tanker pumps ballast water as a tugboat guides it into port. *Left: blickwinkel / Alamy. Right: Islandstock / Alamy*

mussels have had a major impact in their new range, changing energy flows, and causing algal blooms, botulism in birds and declines in fish and mussel populations to the point of local extinctions.[31]

Here in New Zealand, one study in the late 1990s estimated that between 69 and 90% of known non-indigenous marine species are likely to have been introduced via vessel biofouling. Research in the early 2000s found that most vessel types entering New Zealand are likely to have some form of biofouling. Non-native species were found on ~60% of those vessels, and more than 30% of vessels had biofouling species that were new to New Zealand.[32]

The International Maritime Organization (IMO) developed and adopted the International Convention for the Control and Management of Ships' Ballast Water and Sediments (BWM) in 2004, although this did not formally come into force until 2017. The treatment of ballast water is now internationally required. MPI conducted a risk analysis, implementing policy and regulations that align

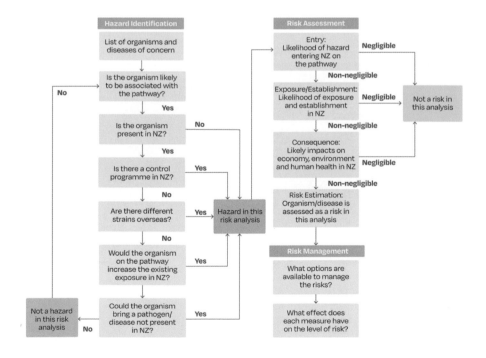

Fig. 2.3: An outline of the risk analysis performed by the Ministry for Primary Industries in New Zealand. This risk analysis builds on the IPPC's framework for pest risk analysis (Fig. 2.1). *Adapted from Georgiades et al. (2020)*[32]

with this convention (Fig. 2.3). The outcome for the risk analysis was to confirm biofouling as a major biosecurity issue for the marine environment. It culminated in the *Craft Risk Management Standard for Biofouling on Vessels Arriving to New Zealand* (CRMS-BIOFOUL). Any ship arriving in New Zealand must now provide evidence that it is compliant with this standard and has a 'clean hull'. The first time a ship was denied entry to New Zealand because of biofouling was in 2017. The DL *Marigold* arrived in Tauranga from Indonesia carrying palm kernel. Divers who inspected the ship's hull found dense fouling of barnacles and tube worms. The boat was ordered to leave New Zealand, so it headed to Fiji for cleaning, although it was also denied entry there because of similar concerns. At the time it was described as the 'world's first biofouling casualty' and 'a benchmark for trading vessels, their owner/managers and charterers'.[33]

## Public Health Emergency of International Concern (PHEIC)

The International Sanitary Conferences began in 1851 with a focus on human health and eventually led to the GATT, the WTO and the SPS Agreement. Despite the increased emphasis on biosecurity, human health has not been forgotten. After its formation in 1948, the WHO issued its first infectious disease prevention regulations in 1951, with a focus on six diseases: cholera, plague, relapsing fever, smallpox, typhoid and yellow fever.

A revised and consolidated version of these regulations, the International Health Regulations (IHR), was adopted by the Health World Assembly in 1969, with revisions and implementation in 2007. The purpose and scope of these are 'to prevent, protect against, control and provide a public health response to the international spread of disease in ways that are commensurate with and restricted to public health risks, and which avoid unnecessary interference with international traffic and trade'.[34] The WHO plays a major role, as it is the agency to which member states are obliged to report any events that may constitute a Public Health Emergency of International Concern (a PHEIC). A PHEIC is 'an extraordinary event which is determined to constitute a public health risk to other States through the international spread of disease and to potentially require a coordinated international response'.[35] WHO member states use an algorithm to determine if a PHEIC exists. Specifically, the WHO should be notified if any two of the four following questions are affirmed:

Is the public health impact of the event serious?

Is the event unusual or unexpected?

Is there a significant risk of international spread?

Is there a significant risk of international travel or trade restrictions?

There are six phases of a PHEIC alert, designed to help countries prepare for and respond to a pandemic. These phases range from discovering a new pathogen or public health concern to the highest phase of observation of widespread and sustained disease transmission among humans. Member states have an obligation under international law to respond promptly to any PHEIC declaration.

The WHO has decreed six PHEIC events since 2005: the 2009 H1N1 pandemic (or 'Swine flu' – see 'Case study 1: The 2009 H1N1 pandemic and the precautionary principle', page 70), the 2014 polio declaration, the outbreak of Ebola in Western Africa in 2014, the 2015–16 Zika virus pandemic, the 2018–20 Ebola epidemic in Kivu, and finally the COVID-19 outbreak in January 2020. PHEIC events are not limited to infectious diseases and may be pronounced after widespread chemical or radioactive pollution events. Five of these six pandemics are almost certainly viral zoonoses that highlight the role of biodiversity in global disease burdens.

The Zika virus pandemic is an example of a zoonosis and PHEIC event. The effects of the virus on pregnant women and their babies has attracted a substantial amount of attention. Zika infects the brain and neural tissue of developing embryos, altering development or causing cell death. These effects can cause microcephaly, where a baby's head is much smaller than expected, and other severe foetal brain abnormalities. The outbreak of clusters of microcephaly in the Americas prompted the PHEIC declaration for Zika virus in 2016.

The very first isolation of the Zika virus was in 1947 from primates in Uganda, West Africa. Evidence suggests that at that time the virus was confined to the equatorial regions of Africa and Asia. Its infection cycle in this region was with monkeys, arboreal mosquitoes and occasionally humans. It was transmitted to humans by bites from infected mosquitoes in the *Aedes* genus, with the widespread invasive species *Aedes aegypti* as a highly competent and anthropophilic (human-loving) vector species. Zika has now spread widely and has been found in Asia, in the Pacific Islands and throughout the tropical regions. Spill-over of the virus quickly appeared in the biodiversity of these recipient communities. Animals that have been found to be infected include the Old and New World primates, elephants, lions, rats, reptiles including snakes, deer, buffalo, bats and sheep.[36]

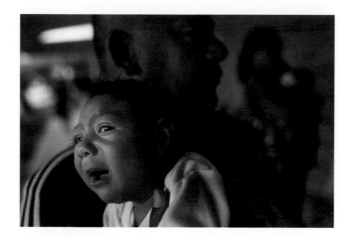

A father holds his one-month-old daughter in Recife, Brazil, 2016. The baby suffers from microcephaly, which manifests in an unusually small head circumference and impaired brain function. The Zika virus, transmitted by mosquitos to pregnant women, is a suspected cause of microcephaly.

*dpa picture alliance / Alamy*

The continuing presence of the virus in these animal communities represents a reservoir pool for the virus, meaning that future local outbreaks and epidemics are highly likely.

What led to the Zika pandemic of 2015–16? David Baud and colleagues wrote in their 2017 review:

> Economic growth in tropical developing countries was a major driver of unprecedented and unplanned urban growth, which provided the ideal ecological conditions for increased Aedes mosquito populations living in intimate contact with crowded human populations. This, combined with ineffective mosquito control and modern transportation, provided the ideal mechanism to transport both the mosquitoes and viruses around the world. As with dengue virus, the resulting increased transmission expanded the probability of genetic change in Zika virus, and thus led to the emergence of viruses with greater epidemic potential and virulence. Therefore, globalisation facilitated the geographical spread of these new viral strains.[37]

The end of the Zika virus PHEIC declaration came in 2016, the same year it began. Case numbers plummeted in both South and North America. This drop was expected by many epidemiologists, because several other mosquito-borne flaviviruses have exhibited similar boom-and-bust dynamics as populations develop herd immunity.[38] Herd immunity is a form of indirect protection from infectious disease when a sufficient percentage of a population has become immune to an infection.

Developing countries face a challenging battle with pandemics. They often have high population densities in warm tropical climates that contain high biodiversity including many invasive species. Access to modern medicine is also an issue, as these countries are unable to afford antiviral drugs that are often stockpiled and hoarded by richer countries when pandemics hit.[39] A highly relevant controversy (and ongoing issue) for developing countries arose during the Avian influenza A (H5N1) or 'bird flu' outbreak of 2007. Avian influenza viruses do not normally infect humans, but this particular strain did, spreading to 17 countries with hotspots in Asia and Africa. A total of 455 people died from this virus over the period of 2003–20. Indonesia experienced the highest mortality of any country. Indonesian officials were particularly concerned about their epidemic, which culminated in their withdrawal from the international flu-sharing system (a global network of influenza laboratories that conduct surveillance of influenza occurrence and strains, assessing the risk of pandemic influenza and also assisting in preparedness). They would not allow genetic sequences and information to be provided to the international community unless there was a guarantee that Indonesia would be afforded the benefits from vaccine development. Vaccine developers and producers are typically located in rich countries, and produce vaccines for profit from sales to many of the same rich countries. Developing countries such as Indonesia are much lower than first on the list for vaccine distribution and deployment. Indonesian officials also argued that researchers in other countries would be taking out patents for vaccines and profiting from the Indonesian genetic data. This stance is legitimate and is recognised in the Nagoya Protocol of the CBD (Table 2.1), which stipulates that a fair and equitable sharing of benefits should arise out of the utilisation of genetic resources. The WHO decreed the Indonesian response as a 'threat to global health security' but as a result of negotiations was asked to take measures that would ensure Indonesia and other developing countries could produce and distribute their own vaccine if a pandemic occured.[40]

Unfortunately, these negotiations have amounted to little and have left almost no legacy. During the current COVID-19 pandemic, the rich, developed countries are being vaccinated while the majority of the poorer developing countries suffer. In March 2021, the *New York Times* reported that 86% of all vaccination shots had gone into arms in high- and upper-middle-income countries. Only 0.1% percent of doses had been administered in low-income countries.[41]

## Summary

The recent evaluations of our current efforts to control pandemics, conserve biodiversity and manage invasive species, disease and pests make for some depressing reading.

The desperate need for effective biosecurity has been highlighted by the SARS-CoV-2 pandemic. Globally, the biosecurity and management of this virus and the COVID-19 disease can only be described as poor. Worse, we are in a 'pandemic era'.[42] The frequency of pandemics is increasing sharply, driven by a growing incidence of emerging diseases. One estimate suggested that there are more than five new diseases emerging in people every year, each with the potential to develop into pandemic proportions. The majority (70%) of these emerging diseases are zoonoses, due to a spillover of microbes from animals after repeated contact between wildlife, livestock and people. While pandemics have their origins in the microbes carried by animal reservoirs, their emergence is entirely driven by human population growth and activities. Pandemics are caused by the same global environmental changes that drive biodiversity loss and climate change. As stated in the 2020 Workshop Report on Biodiversity and Pandemics of the Intergovernmental Platform on Biodiversity and Ecosystem Services (IPBES): 'Without preventative strategies, pandemics will emerge more often, spread more rapidly, kill more people, and affect the global economy with more devastating impact than ever before.'[43] The incidence of WHO PHEIC declarations is only going to increase in future years.

The frequency of biological invasions involving the movement of invasive species across borders is also sharply increasing (Fig. 1.3).[44] This global surge of biological invasions is thought to reflect new trade networks into new regions, and environmental change, resulting in an expanding pool of potential invasive species.[43] Effective international regulation and improved multinational cooperation is crucial to reducing the impact of invasive species on biodiversity, human health and economies.

How have our international efforts and programmes to reduce biological invasions and conserve biodiversity gone so far? Our efforts to achieve the Aichi Biodiversity Targets to reduce the impact of biological invasions and rate of extinctions have fallen well short. Climate change, another international issue, will also increase pest risk in agricultural ecosystems.

A clear conclusion is that the current international regulations are ineffective for both the increasing pandemic problem and the global increase in biological invasions. The existing legislative framework is insufficient for dealing with these growing problems. New approaches have been suggested. In 2007, the 'One Health' concept was formed with the goal of aligning human health and veterinary sciences. The One Health Commission defines this approach as a collaborative, multisectoral and transdisciplinary approach. It works at local, regional, national and global levels to achieve optimal health and well-being outcomes, with the goal of recognising the interconnections between people, animals, plants and their shared environment. Philip Hulme from Lincoln University in Canterbury argues that the One Health concept could be enhanced to One Biosecurity. One Biosecurity would be interdisciplinary, mitigating the effects of invasive species using risk-management tools that extend beyond national borders. A stronger international regulatory framework would be needed on a global scale, involving the new multilateral biosecurity conventions that are responsible for biosecurity governance. Hulme argues for a new International Biosecurity Convention that would be more proactive in reducing the risks of biosecurity threats. Its activities could include establishing and running a global surveillance and monitoring network to provide early warning of new threats. The convention could implement standardised risk-assessment tools and coordinate the development of game-changing management techniques (such as gene editing and microbial biocontrol) to combat invasive pest species of global significance.[45]

Perhaps major benefits would be delivered by a One Biosecurity approach and a new International Biosecurity Convention. It is clear, however, that as Hulme writes: 'The current legislative and scientific tools targeting biological invasions are insufficient to deal with this growing threat and require a new mindset that focuses on curbing the pandemic risk posed by alien species.'[46]

## Case study 1: The 2009 H1N1 pandemic and the precautionary principle

The first Public Health Emergency of International Concern (PHEIC) was the 2009 H1N1 pandemic caused by the H1N1pdm09 virus. The disease was referred to as the H1N1 flu, Swine flu and, unfortunately, Mexican flu, which resulted in Latino communities being stigmatised. The earliest known case was traced to a five-year-old boy in the small rural town of La Gloria, Mexico, which has a massive pig-farming operation nearby that raised around a million pigs in 2008.[47]

Researchers describe the virus as evolving from a mixture or recombination of a North American swine virus that jumped between birds, humans and pigs, and a second swine virus of Eurasian origin that circulated for more than a decade in Mexican pigs before infecting humans. The inter-hemispheric movement of live pigs enabled divergent viral strains to meet, converge and reassort. This new viral strain was of major concern, because the only other known H1N1 pandemic flu was the Spanish flu of 1918, which killed 50–100 million people – the equivalent of 3–5% of the world's population at the time.

The 2009 H1N1 global pandemic had officially claimed fewer than 19,000 people when it ended in 2010. It is likely that this is an underestimate by an order of magnitude, but this death rate was far from the predicted doomsday scenario for a 1918-like pandemic.[47] The estimated case fatality ratio (number of reported deaths per number of reported cases) has been estimated at 0.6% or much lower, which is substantially lower than the 2–3% from the 1918 Spanish flu. Estimates of the viral transmissibility ($R_0$) were calculated at 1.4–1.6, much lower than previous pandemics but slightly higher than seasonal influenza.[48] However, because the WHO decreed the 2009 H1N1 pandemic a 'Phase 6' situation or 'a pandemic in progress', countries bound by the International Health Regulations (2005) were forced under international law to comply with pandemic vaccine production, and the planning and implementation of vaccination campaigns. More than US$18 billion was spent, with pharmaceutical companies seen as the primary beneficiaries.[48] This funding was diverted from other major, ongoing health problems.

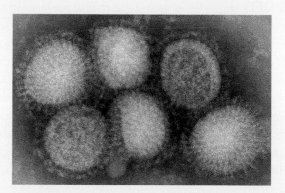

The H1N1pdm09 virus under a microscope.
*Wikimedia Commons*

Substantial debate followed the WHO's handling of the 2009 H1N1 pandemic. Critics noted that the WHO was taking advice on pandemic management from individuals with ties to the pharmaceutical companies. According to a report for the Council of Europe Parliamentary Assembly:

> Some of the outcomes of the pandemic, as illustrated in this report, have been dramatic: distortion of priorities of public health services all over Europe, waste of huge sums of public money, provocation of unjustified fear amongst Europeans, creation of health risks through vaccines and medications which might not have been sufficiently tested before being authorised in fast-track procedures, are all examples of these outcomes.[49]

Mark Davis and colleagues summarised the situation for the WHO, noting:

> Public health appears to be now subject to a mentality of disaster governance that puts it in a precarious double bind: act in case there is a disaster and risk criticism if there is none; fail to act and be criticised if there is a disaster.[50]

Health practitioners supported a cautious approach, saying that, as with any pandemic planning, 'the most predictable thing about flu is unpredictability', and 'Well, we had to do what we thought was right at the time' because 'nobody knew how the virus was going to behave and it could have been catastrophic.' Flu has a substantial level of 'radical uncertainty'. The UK's Health Protection Agency supported the WHO's response, concluding that a precautionary approach was justified to achieve a reduction in viral transmission.

The precautionary principle is based on expectations of technical uncertainty and certainty. If we accept its use in pandemic management, we must also accept that there will be occasions when its application will be excessive and very expensive. Perhaps, however, as some suggest, the use of the precautionary principle will not necessarily lead to wiser decisions in public health policy – and may even be dangerous.[51] Similarly, scientists in New Zealand suggest that the use of the precautionary principle might stifle technological advances in agriculture and biodiversity management.[52]

## *Case study 2: The WTO, the SPS Agreement and NZ exporting apples to Australia*

New Zealand apples have been banned from Australia since 1921. Australia justified the ban on phytosanitary grounds, specifically that New Zealand has the bacterial disease fire blight in apple orchards. Fire blight, caused by the bacterial pathogen *Erwinia amylovora*, can kill blossoms, tree limbs and sometimes entire trees. It gained its common name from the scorched look of infected leaves and branch tips. It is of North American origin but spread to New Zealand by 1919 and Europe in the 1950s. New Zealand requested access to Australian markets in 1986, 1989, 1995 and 1999,[53] but Australian authorities denied these requests and published an import risk assessment in 2006 using semi-quantitative and qualitative methods. New Zealand requested consultation with Australia in 2007, which did not result in a dispute resolution.

As per the process in the SPS agreement, a special panel of the WTO was then formed to consider the issue ('Australia – Apples').[54] Australia included in their WTO submission two other pests of apples present in New Zealand: the fungus European canker (*Neonectria galligena*) and an insect called the apple leafcurling midge (*Dasineura mali*).

The WTO had previous experience with fire blight and apple exports. Fire blight was also the subject of an embargo of apple imports into Japan from

A hawthorn tree infected with fire blight. Fire blight can infect apples, pears, quinces, raspberries and other fruit or ornamental plants.
*Joe / Alamy*

North America ('Japan–Apples').[55] Japan claimed to be free of fire blight, and demanded drastic phytosanitary measures of North American importers: apple orchards must be bordered by a 500-yard scrub brush; three times a year the company must pay the United States Department of Agriculture (USDA) to inspect every tree in the export zones as well as in the buffer zone; and growers must pay to fly in and house Japanese inspectors. The apples also had to be dunked in high concentrations of chlorine, which damaged both apples and machinery.[56] The United States argued and won their case before the WTO. The WTO panel acknowledged that there was considerable evidence that apples did *not* transmit fire blight. The Japanese risk analysis and compliance measures were not based on sufficient scientific evidence. Worse, there was evidence that fire blight currently existed in Japan.

New Zealand authorities similarly criticised the Australian risk assessment for all three pests on their export apples, which the WTO found were not based on solid scientific evidence. The risks stated by Australia had been substantially overestimated. Consequently, Australia's quarantine measures were found to be inconsistent with its obligations under Articles 5.1, 5.2 (relating to risk assessments and analysis) and 2.2 (relating to the use of scientific evidence) of the SPS Agreement. The WTO ruling meant that the Australian market would be opened to apple imports from North America and Europe. New Zealand apples are now exported to Australia.

After this ruling, Australian apple and pear growers gathered in protest, lighting bonfires of dead apple trees to symbolise the fate that they feared for

An apple tree infected with fire blight.
*Andrey Maximenko / Alamy*

Australia's industries. A New Zealand scientist visiting Australia observed fire blight symptoms on Australian fruit trees. He was reported to have said: 'Those lying, cheating Australians really do have fire blight.'[57]

## Case study 3: The International Plant Protection Convention report on the impacts of climate change on plant pests

Climate change is one of the most important challenges to global biodiversity and to humanity. It will influence international trade flows of agricultural products and will change the infectivity, severity and distribution of pests throughout the world. It represents a major challenge to the international plant health community and its ability to respond in a scientific, decisive and unified manner.

In 2021 the Secretariat of the IPPC released a report on the impact of climate change on plant pests. Below are the key messages.

1. Climate change increases pest risks in agricultural ecosystems, especially in cooler Arctic, boreal, temperate and subtropical regions. Some pests have already expanded their host range or distribution due to changes in climate.

2. Increased pest risks may pose a threat to the environment because invasive pests are one of the main drivers of biodiversity loss.

3. Weather is the second most important factor for pest dispersal after international travel and trade. Temperature, humidity, light, wind and any combination of these can influence the life cycle of pests.

4. Climate change effects on pest species are complex. They may be direct or indirect, and also interact with each other. Possible effects on pests include increased risk of pest introduction and changes to their geographical distribution, seasonal phenology and population dynamics. The effects are easier to predict for pest species that are mainly affected by temperature. Prediction is more difficult for pests whose reproduction and dispersal are strongly related to water availability, wind, and crop management.

5. Prevention is the most efficient and effective strategy to minimise the impact of a new pest. Climate change considerations should be included in a country or region's management of phytosanitary regulatory systems.

6. It is necessary to adjust plant protection methods to respond to the impact of climate change. Maintaining ecosystem services under climate change conditions is key to protecting plant health, sustaining the environment

and ensuring food security.

7.   Pests do not respect borders. International cooperation is critical to ensure that countries can adapt their pest risk management measures to climate change. It is important to examine how cooperation can enhance effective pest management and allow for the development of harmonised strategies.

8.   It is important to intensify national, regional and international surveillance and monitoring activities for plant health threats. Multilateral surveillance programmes should be enhanced to offset phytosanitary threats.

9.   Pest risk analysis activities need to be intensified at national, regional and international levels, and climate change considerations should be included in the assessment of pest risks.

10.  Policymakers should encourage countries to conduct phytosanitary capacity evaluations with the support of the IPPC Secretariat. This will result in greater national phytosanitary capacity and cost-benefit improvements.

11.  To protect plant health worldwide, policymakers should encourage the use of environmentally friendly methods. These include integrated pest management, strengthening the phytosanitary capacity of plant protection organisations, and supporting coordinated responses among countries. An active, official international information exchange mechanism dedicated to providing data about the occurrence and risk of pests and the development of potential pathways should be established.

12.  Multidisciplinary approaches and collaboration are beneficial when assessing and managing the impact of climate change on pests, and consequently on plant health. It is important to enhance knowledge-sharing among inter alia, plant pathologists, entomologists, meteorologists, weed scientists, agronomists and microbiologists. It would be beneficial to strengthen cooperation among experts working on human, animal and environmental health in various ecosystems and sectors, such as agriculture, forestry and unmanaged ecosystems (see the Circular Health or One Health approaches).

13.  To better inform policymaking with up-to-date scientific data, policymakers should support empirical research on the impact of climate change on pests and, by extension, on plant health. Establishing a global mechanism for research coordination would help international efforts to protect agriculture, the environment and trade activities from pests.

One of the examples in the IPPC report is the potato late blight (*Phytophthora infestans*), as discussed in Chapter 1. The potential for the increasing survival of oomycetes in a polewards direction as a result of climate change presents a challenge for potato protection and production. Favourable conditions in winter will likely enable a build-up of pathogen inoculum on early cultivars at the start of the growing season, increasing pathogen pressure throughout the season. Consequently, climate change may lead to late blight epidemics.[22]

## Further reading and discussion

1. The highly pathogenic H5N1 avian influenza emerged in Hong Kong. By 2006 it had caused outbreaks in poultry or wild birds in 53 countries, with at least 256 human cases, including 151 deaths. What role did natural bird migration have in the international transmission of this disease compared with transmission via international trade? What management lessons can we learn from differences in disease transmission between regions and countries?

*See:* Kilpatrick, A. M., et al. (2006). Predicting the global spread of H5N1 avian influenza. *Proceedings of the National Academy of Sciences 103*(51), 19368–19373. doi.org/10.1073/pnas.0609227103

Deforestation in the Tasman District of the South Island, New Zealand. *Martin Wegman / Wikimedia Commons*

2. New Zealand is a signatory to the Convention on Biological Diversity and the associated international obligations on the Aichi Biodiversity Targets. How would you rate New Zealand's approach and work towards these targets? What would you do differently?

*See:* Department of Conservation (2019). *New Zealand's Sixth National Report to the United Nations Convention on Biological Diversity. Reporting period: 2014–2018.* Wellington, New Zealand, Department of Conservation, 124. cbd.int/doc/nr/nr-06/nz-nr-06-en.pdf

3. In a review of the global response to biological invasions, Phil Hulme from Lincoln University concluded: 'The current legislative and scientific tools targeting biological invasions are insufficient to deal with this growing threat and require a new mindset that focuses on curbing the pandemic risk posed by alien species.' He recommends 'One Biosecurity' as a new approach. What is One Biosecurity, and would it help? What else would you do?

*See:* Hulme, P. E. (2021). Unwelcome exchange: International trade as a direct and indirect driver of biological invasions worldwide. *One Earth 4*(5), 666–679. doi.org/10.1016/j.oneear.2021.04.015

4. How would investment that prevents deforestation and regulates wildlife trade limit zoonoses and future pandemics? Is it possible to convince international agencies that preventive efforts would be substantially less than the economic and mortality costs of responding to these pathogens once they have emerged?

*See:* Dobson, A. P., et al. (2020). Ecology and economics for pandemic prevention. *Science 369*(6502), 379–381. doi.org/10.1126/science.abc3189

# References

1. Druett, J. (1983). *Exotic intruders: The introduction of plants and animals into New Zealand*. Heinemann.

2. Sadler, R. (2021, March 9). Snake found on construction site in Auckland. *Newshub*. newshub.co.nz/home/new-zealand/2021/03/snake-found-on-construction-site-in-auckland.html

3. Biosecurity Act 1993. legislation.govt.nz/act/public/1993/0095/latest/whole.html

4. Food and Agriculture Organization of the United Nations. (2021, March 15). *Biosecurity*. fao.org/food/food-safety-quality/a-z-index/biosecurity/en/

5. Bianchine, P. J., & Russo, T. A. (1992). The role of epidemic infectious diseases in the discovery of America. *Allergy and Asthma Proceedings, 13*(5), 225–232. doi.org/10.2500/108854192778817040

6. Fenn, E. A. (2000). Biological warfare in eighteenth-century North America: Beyond Jeffery Amherst. *The Journal of American History, 86*(4), 1552–1580. ncbi.nlm.nih.gov/pubmed/18271127

7. Atkinson, U. A. E. (1973). Spread of the ship rat (*Rattus r. rattus* L.) in New Zealand. *Journal of the Royal Society of New Zealand, 3*(3), 457–472. doi.org/10.1080/03036758.1973.10421869

8. Roser, M. (2017). *Tourism*. Our World in Data. ourworldindata.org/tourism

9. International Chamber of Shipping. (2021, July 26). *Shipping and world trade: Driving prosperity*. ics-shipping.org/shipping-fact/shipping-and-world-trade-driving-prosperity/

10. Frith, J. (2012). The history of plague – Part 1. The three great pandemics. *Journal of Military and Veterans' Health, 20*(2), 11–16.

11. Seebens, H., Blackburn, T. M., Dyer, E. E., et al. (2018). Global rise in emerging alien species results from increased accessibility of new source pools. *Proceedings of the National Academy of Sciences, 115*(10), E2264–E2273. doi.org/10.1073/pnas.1719429115

12. Fidler, D. P. (2005). From international sanitary conventions to global health security: The new International Health Regulations. *Chinese Journal of International Law, 4*(2), 325–392. doi.org/10.1093/chinesejil/jmi029

13. Daswani, A. G., Santana, R., & Volkai, J. N. (2018). *GATT disputes: 1948–1995*. World Trade Organization. wto.org/english/res_e/publications_e/gatt4895vol12_e.htm

14. World Trade Organization. (2021, April 12). The WTO Agreement on the Application of Sanitary and Phytosanitary Measures (SPS Agreement). wto.org/english/tratop_e/sps_e/spsagr_e.htm

15. World Trade Organization. (2000). DS18: Australia – Measures affecting importation

of salmon. wto.org/english/tratop_e/dispu_e/cases_e/ds18_e.htm

16. World Organisation for Animal Health. (2022, January 31). Who we are. oie.int/en/who-we-are/

17. World Organisation for Animal Health. (2021, July 6). Animal diseases. oie.int/en/what-we-do/animal-health-and-welfare/animal-diseases/

18. Lacey, M. (2009, April 28). From Édgar, 5, coughs heard around the world. *New York Times*.

19. IPPC Secretariat. (2017). *Determination of pest status in an area: International Standard for Phytosanitary Measures 8*. Food and Agriculture Organization of the United Nations. ippc.int/static/media/files/publication/en/2017/06/ISPM_08_1998_En_2017-05-23_PostCPM12_InkAm.pdf

20. IPPC Secretariat. (2016). *Framework for pest risk analysis: International Standard for Phytosanitary Measures 2*. Food and Agriculture Organization of the United Nations. ippc.int/static/media/files/publication/en/2016/01/ISPM_02_2007_En_2015-12-22_PostCPM10_InkAmReformatted.pdf

21. IPPC Secretariat. (2016). *Guidelines for pest eradication programmes: International Standard for Phytosanitary Measures 9*. Food and Agriculture Organization of the United Nations. ippc.int/static/media/files/publication/en/2016/01/ISPM_09_1998_En_2015-12-22_PostCPM10_InkAmReformatted.pdf

22. IPPC Secretariat. (2021). *Summary for policymakers of the scientific review of the impact of climate change on plant pests: A global challenge to prevent and mitigate plant pest risks in agriculture, forestry and ecosystems*. Food and Agriculture Organization of the United Nations. fao.org/documents/card/en/c/cb4777en/

23. World Health Organization & the Food And Agriculture Organization of the United Nations. (2005). *Understanding the Codex Alimentarius: Revised and updated*. fao.org/3/y7867e/y7867e00.htm

24. Rothschild, M. (2013, June 3). Codex Alimentarius: Book of food or book of death? *Skeptoid Media*. skeptoid.com/blog/2013/06/03/codex-alimentarius-book-of-food-or-book-of-death/

25. Secretariat of the Convention on Biological Diversity. (2020). *Global Biodiversity Outlook 5*. cbd.int/gbo/gbo5/publication/gbo-5-en.pdf

26. Jones, H. P., Holmes, N. D., Butchart, S. H. M., et al. (2016, April 12). Invasive mammal eradication on islands results in substantial conservation gains. *Proceedings of the National Academy of Sciences, 113*(15), 4033–4038. doi.org/10.1073/pnas.1521179113

27. Department of Conservation. (2019). *New Zealand's sixth national report to the United Nations Convention on Biological Diversity. Reporting period: 2014–2018*. doc.govt.nz/globalassets/documents/about-doc/role/international/nz-6th-national-report-convention-biological-diversity.pdf

28. Drake, J. M., & Lodge, D. M. (2007). Hull fouling is a risk factor for intercontinental

species exchange in aquatic ecosystems. *Aquatic Invasions, 2*(2), 121–131. doi.org/10.3391/ai.2007.2.2.7

29. Havel, J. E., Kovalenko, K. E., Thomaz, S. M., et al. (2015, January 25). Aquatic invasive species: Challenges for the future. *Hydrobiologia, 750*(1), 147–170. doi.org/10.1007/s10750-014-2166-0

30. Connelly, N. A., O'Neill, C. R., Jr., Knuth, B. A., et al. (2007). Economic impacts of zebra mussels on drinking water treatment and electric power generation facilities. *Environmental Management, 40*, 105–112. doi.org/10.1007/s00267-006-0296-5

31. Higgins, S. N., & Vander Zanden, M. J. (2010, May). What a difference a species makes: A meta-analysis of dreissenid mussel impacts on freshwater ecosystems. *Ecological Monographs, 80*(2), 179–196. doi.org/10.1890/09-1249.1

32. Georgiades, E., Kluza, D., Bates, T., et al. (2020, June 19). Regulating vessel biofouling to support New Zealand's marine biosecurity system – A blueprint for evidence-based decision making. *Frontiers in Marine Science, 7*, 390. doi.org/10.3389/fmars.2020.00390

33. The Maritime Executive. (2017, March 9). *Biofouling: Ship refused entry to New Zealand, Fiji.* maritime-executive.com/article/biofouling-ship-refused-entry to-new-zealand-fiji

34. World Health Organization. (2008). *International health regulations (2005)* (2nd ed.). who.int/publications/i/item/9789241580410

35. World Health Organization. (2019). *Emergencies: International health regulations and emergency committees.* who.int/news-room/questions-and-answers/item/emergencies-international-health-regulations-and-emergency-committees

36. Bueno, M. G., Martinez, N., Abdalla, L., et al. (2016). Animals in the Zika virus life cycle: What to expect from megadiverse Latin American countries. *PLOS Neglected Tropical Diseases, 10*(12), e0005073. doi.org/10.1371/journal.pntd.0005073

37. Baud, D., Gubler, D. J., Schaub, B., et al. (2017). An update on Zika virus infection. *The Lancet, 390*(10107), 2099–2109. doi.org/10.1016/S0140-6736(17)31450-2

38. Cohen, J. (2017, August 16). Zika has all but disappeared in the Americas. Why? American Association for the Advancement of Science. sciencemag.org/news/2017/08/zika-has-all-disappeared-americas-why

39. Vilhelmsson, A., & Mulinari, S. (2018). Pharmaceutical lobbying and pandemic stockpiling of Tamiflu: A qualitative study of arguments and tactics. *Journal of Public Health, 40*(3), 646–651. doi.org/10.1093/pubmed/fdx101

40. Enserink, M. (2007). Avian influenza: Indonesia earns flu accord at World Health Assembly. *Science, 316*(5828), 1108. doi.org/10.1126/science.316.5828.1108

41. Collings, K., & Holder, J. (2021, April 1). See how rich countries got to the front of the vaccine line. *New York Times.*

42. The Lancet Planetary Health. (2021). A pandemic era. *The Lancet Planetary Health, 5*(1), e1. doi.org/10.1016/S2542-5196(20)30305-3

43. Daszak, P., Amuasi, J., das Neves, C. G., et al. (2020). *IPBES workshop on biodiversity and pandemics: Workshop report*. Intergovernmental Science-Policy Platform on Biodiversity and Ecosystem Services. ipbes.net/sites/default/files/2020-12/IPBES%20Workshop%20on%20Biodiversity%20and%20Pandemics%20Report_0.pdf

44. Pyšek, P., Hulme, P. E., Simberloff, D., et al. (2020). Scientists' warning on invasive alien species. *Biological Reviews, 95*(6), 1511–1534. doi.org/10.1111/brv.12627

45. Hulme, P. E. (2021). Advancing One Biosecurity to address the pandemic risks of biological invasions. *BioScience, 71*(7), 708–721. doi.org/10.1093/biosci/biab019

46. Hulme, P. E. (2021). Unwelcome exchange: International trade as a direct and indirect driver of biological invasions worldwide. *One Earth, 4*(5), 666–679. doi.org/10.1016/j.oneear.2021.04.015

47. Viboud, C. & Simonsen, L. (2012). Global mortality of 2009 pandemic influenza A H1N1. *The Lancet Infectious Diseases, 12* (9), 651–653. doi.org/10.1016/S1473-3099(12)70152-4

48. Pada, S. & Tambyah, P.A. (2011). Overview/reflections on the 2009 H1N1 pandemic. *Microbes and Infection* 13, 470–478.

49. Cohen, D. & Carter, P. (2010). Conflicts of interest: WHO and the pandemic flu 'conspiracies'. *The British Medical Journal, 340*, c2912.

50. Davis, M., Flowers, P. & Stephenson, N. (2014). 'We had to do what we thought was right at the time': Retrospective discourse on the 2009 H1N1 pandemic in the UK. *Sociology of Health & Illness, 36*(3), 369–82. doi.org/10.1111/1467-9566.12056

51. Gignon, M., Ganry, O., Jarde, O. & Manaouil, C. (2013). The precautionary principle: Is it safe. *European Journal of Health Law, 20* (3), 261–70. doi.org/10.1163/15718093-12341272

52. Goldson, S.L., Frampton, E.R. & Ridley, G.S. (2010). The effects of legislation and policy in New Zealand and Australia on biosecurity and arthropod biological control research and development. *Biological Control, 52* (3), 241–244. doi.org/10.1016/j.biocontrol.2009.03.006

53. Arcuri, A., Gruszczynski, L. & Herwig, A. (2010). Risky apples again? Australia – measures affecting the importation of apples from New Zealand. *European Journal of Risk Regulation 1,* 437–443. doi.org/10.1017/S1867299X00000933

54. World Trade Organization (2011). *DS367: Australia – Measures affecting the importation of apples from New Zealand.*

55. World Trade Organization (2005). *DS245: Japan – Measures affecting the importation of apples.*

56. Helm, L. & Eisenstodt, G. (1996, July 22). Caught in cross-fire of Pacific apple war. *Los Angeles Times.*

57. Carey, A. (2011, July 29). NZ apples will burn us: Orchardists. *The Age.*

# 3. PEST CONTROL LEGISLATION AND LAW IN NEW ZEALAND

## How the government is equipped to deal with pests and potential pests

Pests new to New Zealand are found every year. We might find an exotic fruit fly in an orchard. Perhaps we discover the virus for foot-and-mouth disease in a dairy herd. Despite our best efforts, exotic organisms are regularly landing at the borders of all countries. And some, after breaching these borders, will become established and widespread. German wasps, for example, are now well established throughout New Zealand, stinging thousands since their introduction in the mid-1900s and occasionally killing members of the public. Adding to the pest burden, some native species will also become economically costly, a health issue, or a nuisance deemed to require management. What happens next? How are the government and government bodies legislatively enabled to manage pests?

In this chapter I'll examine New Zealand's legislative regime governing pest management, with the primary focus on exotic pests and pathogens of animals. There are two key pieces of legislation for the management of these species that I'll cover extensively here. The first is the Biosecurity Act 1993, and the second is the Hazardous Substances and New Organisms (HSNO) Act 1996.

The purpose of the Biosecurity Act 1993 is to enable the exclusion, eradication and effective management of pests and unwanted organisms.[1] The legislation covers pre-border risk management, border management for quarantine, readiness and response, and long-term pest management. The first role of this act is to keep exotic species out, by monitoring them and placing restrictions on what can and cannot be imported. Should something breach the border, the next goal is to enable officials to act quickly to find and kill any newly arrived incursions. If a pest or an unwanted organism does become established, there are two pest control plans implemented under this legislation. The first approach to dealing with a pest is a 'nuclear' option – a national pest management plan. These plans are a big deal: currently only three of these exist in New Zealand. National pest management plans are typically considered for nationally distributed pests that generate substantial economic costs. Bovine tuberculosis is one example: it is a nationwide pest with major economic impact on New Zealand's beef exports.

Many more pests are managed on local or regional scales. Regional pest management plans attempt to locally limit the abundance and effects of pests. Region-specific approaches are logical, for example, when a pest has a small distribution in a particular climate and is unlikely to spread throughout the entire nation, such as German wasps, which are a major pest in many New Zealand regions but not in others.

The second key piece of legislation for biosecurity in New Zealand is the HSNO Act 1996. Its purpose is to 'protect the environment, and the health and safety of people and communities, by preventing or managing the adverse effects of hazardous substances and new organisms'. This act is about safeguarding the life-supporting capacity of New Zealand's air, water, soil and ecosystems. The sections on 'new organisms' are particularly relevant for biosecurity and pest management. New organisms are typically species that people would like to

A European rabbit (*Oryctolagus cuniculus*).
*JM Ligero Loarte / Wikimedia Commons*

import into the country, perhaps as a new crop, or as a predator or parasite for the purposes of the biological control of pest species. They are defined as those species, or even strains or varieties of species, not in the country before the arbitrary date of 29 July 1998. These new organisms also include genetically modified species. The precautionary approach discussed in the previous chapter is specifically stated in the HSNO Act, where 'all persons exercising functions, powers, and duties under this Act . . . shall take into account the need for caution in managing adverse effects where there is scientific and technical uncertainty about those effects'.[2]

There are other laws and legislation that come into play for biosecurity and pest management. For example, the Wildlife Act 1953[3] and the Wild Animal Control Act 1977 both have a role in determining which species are pests for control and which are precious and in need of conservation. The Wild Animal Control Act 1977 regulates the control of harmful species of introduced wild animals and the operations of recreational and commercial hunters, so as to achieve concerted action and effective wild animal control.[4] The Agricultural Compounds and Veterinary Medicines (ACVM) Act 1997 is important too, because it regulates the use of pesticides or compounds for managing or eradicating pests.[5]

Interestingly, of all the newish legislation, the ACVM Act is the only one to give a definition of a pest. Under the ACVM Act 1997, a pest is defined as 'any unwanted living organism including micro-organisms, pest agents, and any genetic structure that is capable of replicating itself (whether that structure comprises all or only part of an entity, and whether it comprises all or only part of the total genetic structure of an entity) that may affect plants, animals, or raw primary produce; and includes any entity declared to be a pest for the purposes of this Act by Order in Council made under subsection (2)'. Despite any speculation on my or your behalf about many of our fellow New Zealanders, this pest definition does not and cannot include 'any human being or living organism which affects only human beings; and any living organism declared not to be a pest'. In contrast, the Biosecurity Act 1993 somewhat tautologically defines a pest as 'an organism specified as a pest in a pest management plan'.

I'll include some examples and a discussion of acts and legislation where relevant. The use of legislation for the control of human pathogens and disease will be discussed in Chapter 6.

## A team of 4.7 million people monitoring 'Unwanted Organisms'

A recent government initiative for the year 2025 is to make a 'biosecurity team of 4.7 million'.[6] The goal is to have everyone looking out for pests or diseases, so that newly arrived pests or pestilence can be caught and dealt with before they become established and widespread.* It is, however, already the law under the Biosecurity Act 1993 that every New Zealander has a duty to inform the Ministry for Primary Industries (MPI) if they see an organism that is not normally seen or otherwise detected in New Zealand. This means we have an obligation to report any species believed to be not already established here.

The New Zealand government maintains a register of 'Unwanted Organisms', or species that the government believes are capable of causing harm to any natural and physical resources or human health. There were more than 14,800 entries on the government's watchlist or unwanted organisms database in 2020.[7] Of these 8029 are insects, 530 are viruses, 271 are plants or weeds, and 49 are vertebrate species. Some of the species on this database are already well established in New Zealand, including bovine tuberculosis, ferrets and rooks. We are lucky not to have several other species, including the red fox, despite it being proposed for introduction after European colonisation in the 1800s. This database also contains five prions. Prions aren't 'alive', or organisms. Instead, they are a type of protein that can trigger proteins in the brain to fold abnormally. The most well-known disease caused by prions is mad cow disease, or bovine spongiform encephalopathy (BSE). Prions have been seen in sheep in New Zealand and can be devastating for humans too, such as with the Kuru disease among people in Papua New Guinea (see 'Case study 1: Prions, mad cows and sheep, and cannibalism', page 111).

Some of these unwanted organisms are awarded the status of 'Notifiable Organisms', which are pests and diseases that must be reported to MPI if observed in New Zealand.[8] Approximately 270 species are included in this list. The vast majority of entries are those species that affect the economy via the

---

* MPI's pest and disease hotline is 0800 80 99 66. They encourage you to photograph and catch the pest, if safe to do so. Photos can help specialists interpret your findings. When you are photographing a potential pest, MPI encourages you to place it against a white background, include a common object for scale (such as a coin) and use the default settings on your camera. Don't zoom in or add filters. Note that there are some organisms you shouldn't touch: snakes or venomous animals, obviously, but also plant pathogens like the fungal disease myrtle rust (*Austropuccinia psidii*).

agricultural, horticultural or forestry industries. Again, some of these pathogens, plants or animals are present in the country, while others are species New Zealand biosecurity authorities are desperately trying to keep out. (See 'Case study 2: 55,000 fruit trees might be diseased and have to be destroyed', page 113.)

Bovine tuberculosis (*Mycobacterium bovis*), for example, is listed as Unwanted though not as a Notifiable Organism. Another cattle disease on the Notifiable Organisms list is rinderpest, caused by the rinderpest virus, which is an interesting addition to the list because it was declared globally eradicated in 2011. If a New Zealand citizen even suspects the presence of rinderpest (hopefully not, given it has been globally eliminated) or any other Notifiable Organism in the country, it is their legal obligation to report it to MPI.

## 'Oh my God, I've found a fruit fly'

Of the 270 species listed as 'Notifiable Organisms', 31 are fruit flies or fruit fly relatives. While the New Zealand government expects the public to be on the lookout for these flies, they also have an annual National Fruit Fly Surveillance Programme to monitor these pests. Fruit flies are considered worthy of a specific surveillance programme because of their economic importance. The horticultural industry earned New Zealand an estimated $6.2 billion in 2019, with 80% of these earnings coming from plants that are potential fruit-fly hosts.[9] There could be substantial fruit losses in orchards, increased horticultural insecticide use and international trade restrictions if fruit fly were to become established here.

All international visitors entering New Zealand are told about the biosecurity risks of bringing produce such as fresh fruit into the country. Some foods, drinks, sports and outdoor equipment and animal and plant products can carry diseases and pests such as fruit flies. Visitors have to declare if they have any of these goods. Under section 154 of the Biosecurity Act 1993, quarantine officials can search people, their personal belongings and their goods, and impose fines or worse. The penalty for a false declaration is a NZ$400 infringement fee. But those convicted of deliberate smuggling can be fined up to NZ$100,000 and sentenced to up to five years in prison. In 2016, for example, an Australian passenger flew into Wellington from Melbourne. The visitor had a single mandarin, perhaps forgotten somewhere in the depths of their luggage. MPI staff found four Queensland fruit-fly larvae in that single item of fruit. 'One of our quarantine officers detected the fruit when the passenger's bag went through an MPI biosecurity X-ray machine. Another officer discovered insect damage on the mandarin and pulled the skin off, finding the larvae nestled inside.' The passenger had the mandarin confiscated and was handed a NZ$400 fine.[10]

Does that sound like a lot of bother and effort for one mandarin? Well, to put all this in context, how much does it cost for each fly that slips past the border? The discovery of just 14 fruit flies in Auckland over the summer of 2019 ended up with an estimated cost of NZ$18 million to the taxpayer. As a reporter from the *New Zealand Herald* described it, that is just under NZ$1.3 million per fly.[11]

Those 14 flies were discovered by the National Fruit Fly Surveillance Programme. Every year, from spring to winter, this programme is mounted throughout New Zealand. In 2020, a total of 7877 traps were placed throughout the country.[9] Ports and large cities are a focus of the trapping efforts, but

Fig. 3.1: A summary of MPI's response to the incursion of the Queensland fruit fly (*Bactrocera tryoni*) in the North Shore, Auckland, during the summer of 2019–20. The discovery of one male fly in February 2019 initiated the response, resulting in 800 people working on the surveillance and eradication programme. The response was ended almost a year after the initial detection. *MPI*

horticultural regions are also well sampled. Male flies look for females using chemical cues, so the traps contain artificial sex pheromones that lure hopeful males into the containers, where they are killed with insecticides. The collected flies are then submitted for identification. The whole system is checked or undergoes quality control when sneaky researchers plant dead flies in the traps and see how many make it through to identification.

Since 1989 there have been 12 separate detections of economically important fruit flies via this surveillance programme. Nine of these have been in New Zealand's largest city, Auckland. The typical MPI response is a well-oiled machine that implements massive, localised surveillance and trapping. Pheromone traps are deployed and fruit and fruit trees are examined. More often than not, these surveillance programmes do not find further evidence of fruit flies and the programme is ended. Perhaps the trapped male fly slipped into the country on an item of imported fruit. But sometimes, unfortunately, more flies are discovered, and an eradication plan is enacted.

The detection of fruit flies in Auckland in 2019 and the consequent NZ$18 million fruit-fly response by MPI involved a workforce of over 800. These people had a broad range of tasks, including deploying traps, surveying and examining fruit and fruit trees, and identifying flies. There were major costs in the publicity campaign as well, designed to ensure that the general public understood and supported the eradication. Public support is essential (and difficult) when goods and produce are not allowed to be removed from large areas of land under quarantine, or when large amounts – 145 tonnes in this case – of otherwise quality produce is destroyed (Fig. 3.1).

Fruit and vegetable growers are clearly invested in fruit-fly eradications. With the goal of formalising and leveraging this investment, the New Zealand government has also created legislation under the Biosecurity Act 1993 to directly involve the industry in incursion decisions and preparedness. A Government Industry Agreement for Biosecurity Readiness and Response (GIA) is a partnership between government and industry representatives with the goal of improving biosecurity. The signatories negotiate and agree on the priority pests and diseases that are of most concern. They agree on an action plan to minimise the risk and impact of incursions, or to prepare for and manage a response in the event than an incursion occurs. While the industry groups share in the decision-making, they also share in the costs. In a 2015 Queensland fruit-fly response in Auckland, two GIA partners – Kiwifruit Vine Health and New Zealand Apples

& Pears – worked with MPI to decide how to respond to the incursion. They provided resources and technical support to their members. These and other horticultural industry groups have now developed the New Zealand Fruit Fly Strategy 2017–22 to proactively manage future incursions.[12] Their vision is a New Zealand free from economically significant fruit fly. There are now an increasing number of GIAs for a variety of industries, including aquaculture.

The National Fruit Fly Surveillance Programme is just one of many that are run annually around New Zealand. The *Atlas of Biosecurity Surveillance* lists 11 different surveillance programmes coordinated by the government. They include programmes monitoring exotic ants, gypsy moth, mosquitoes and arboviruses (arthropod-borne viruses, or viruses that are transmitted by arthropod vectors), wildlife diseases and avian influenza.[13] I suspect that many people will view surveillance as an excessive expense, until a new pest species establishes, becomes widespread and costs the industries or the public millions in lost produce or pest management.

The ability to detect biological invaders is crucial for an effective surveillance programme, or for pest management and eradication purposes. The biggest challenge is that an effective detection programme requires the identification of the individual *least* likely to be observed. Unfortunately, however, our ability to detect individuals varies between and within species. For fruit fly and gypsy moth, the traps only trap males, because the lure used is an artificial female sex pheromone. For mosquitoes, adult trapping will only find females, because the carbon dioxide lure is only attractive for individuals seeking blood meals for egg production. Even within the sexes there is likely to be variation in detectability.

One of the best analyses of pest detectability comes from an analysis of a mark–recapture study on the invasive brown tree snake (*Boiga irregularis*) in Guam.[14] As they are ineffective for smaller snakes, traps had been ruled out as a detection method for monitoring these invasive pests.[15] Consequently, visual detection methods were one of few options available. The researchers, however, discovered considerable variation in their ability to find snakes in the forests. There was a lower probability of detecting small and large females than males of those sizes. Snakes with poor body condition were more likely to be observed than snakes with high body condition, perhaps because if the snake was hungry, its active hunting behaviour made it more observable. The environmental conditions played a role in detection. Surveys conducted in bright moonlight and with strong wind gusts showed a decreased probability of detecting snakes.

And people played a part too: experienced search teams had twice the ability to detect snakes than did teams with less experience.

This variation in detectability between individuals in a population is likely to occur for any species and can make surveillance ineffective or quash eradication efforts. 'Trap-shy' individuals, for example, appear to occur in many animal populations and have led to false declarations of pest eradication.[16]

## National pest management plans, for widespread and devastating incursions

Exotic or invasive pests have and will become established in New Zealand despite the best efforts of quarantine officials, GIAs and millions of citizens monitoring 'notifiable' or 'unwanted' organisms. The Biosecurity Act 1993 enables the management of these pests by either national pest management plans or regional pest management plans.

National pest management plans are intended to provide for the countrywide coordinated eradication or management of pests. They are typically for major pests of national significance and have statutory force. To be considered for a national plan, pests must be associated with adverse effects on economic well-being; health, soil, water or air quality; the viability of threated species; animal welfare; or the relationship between Māori, their culture, their traditions and their ancestral lands, waters, sites, wāhi tapu (sacred sites) and taonga. Plans must have a management agency coordinating the response, and must provide methods for funding. Rules may require people to monitor the presence or distribution of a pest, or to control or eradicate it, and levies can be imposed.

As of 2021, there are just three national pest management plans in operation. Two of these highlight eradication or elimination of diseases in agricultural animals as objectives. The National Bovine Tuberculosis Pest Management Plan seeks the eradication of *Mycobacterium bovis*, which causes bovine tuberculosis, from New Zealand.[17] The key milestones are the New Zealand-wide freedom from bovine tuberculosis in cattle and deer by 30 June 2026, and freedom from bovine tuberculosis in possums by 30 June 2040. This plan and its progress are discussed in the following chapter on eradication (see 'Case study 2: The eradication programme for bovine tuberculosis in New Zealand' in Chapter 4, page 152). The second is the National American Foulbrood Pest Management

Fig. 3.2. Kiwifruit Vine Health was established in 2010. It is now the lead organisation for managing biosecurity readiness, response and operations on behalf of the kiwifruit industry. This poster lists its most unwanted pests. *KVH*

Plan, which aims to manage American foulbrood (caused by the spore-forming bacteria *Paenibacillus larvae larvae*) in honey bees. The primary objective of this plan is to reduce the reported incidence of American foulbrood by an average of 5% each year.[18] The secondary objectives include eliminating American foulbrood in beehives by destroying any cases and associated bee products.

The third national pest management plan currently in effect is the National Psa-V Pest Management Plan.[19] *Pseudomonas syringae* pv. *actinidiae* (Psa or Psa-V) is a bacterial species that was first identified in New Zealand in November 2010, that causes a disease called bacterial canker of kiwifruit. It is widely believed that it entered the country on imports of pollen that were to be used for artificial insemination and crop improvement.[20] The identification of the bacteria was delayed because the damage symptoms are similar to those of closely related existing bacterial diseases, and this enabled the newly arrived pathogen to become widely established and distributed prior to formal discovery and description.[21] The pathogen rapidly caused widespread and severe impacts on New Zealand's NZ$1 billion kiwifruit industry, including the death of kiwifruit vines. The social and economic effects of the pathogen were considerable. Orchardist suicide was considered likely, and a specific goal to avoid any deaths was set at the very beginning of the incursion.[21] Rather than elimination or eradication, the National Psa-V Pest Management Plan has the primary objective of reducing the harmful effects of the bacteria on economic well-being by preventing its spread and minimising its impact on kiwifruit production. The secondary objectives are about keeping the disease contained and reducing its distribution and effects in regions where infections currently occur. An organisation called Kiwifruit Vine Health was created to manage Psa and signed a GIA for kiwifruit biosecurity in 2014. Kiwifruit Vine Health has become the lead organisation for managing Psa and for potential future incursions (Fig. 3.2).

Very occasionally (well, once to date), the government and industry will admit defeat and end a national pest management plan. For example, in 2000 an incursion of the parasitic mite *Varroa destructor* was found in honey bee colonies in the upper part of the North Island near Auckland. This mite is acknowledged as one of the world's worst pests for honey bees,[22] and it quickly spread through the North Island (Fig. 3.3). The North and South Islands are separated by a substantial oceanic barrier, Cook Strait, which had the potential to form a natural obstacle to mite movement. In 2005 a national pest management plan was put into effect with the specific goal of preventing *Varroa* establishing in the

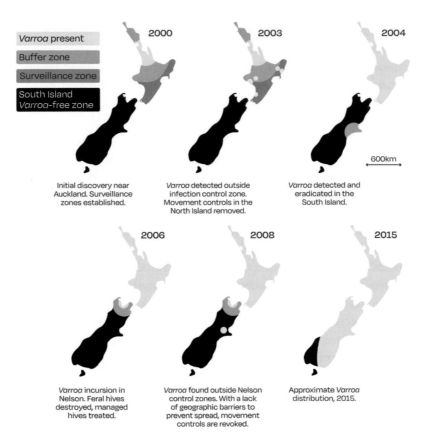

Fig. 3.3: The discovery of *Varroa* and its spread in New Zealand. A national pest management plan was enacted in 2005, with the goal of stopping it from spreading to the South Island, but was revoked in 2012. *Modified from Iwasaki et al. (2015)*[25]

South Island.[23] A new industry group called the Varroa Agency was created to administer the plan, and collected a levy from beekeepers to fund surveillance and movement controls. Small incursions into the South Island were found and appeared to be eradicated. In June 2006, however, *Varroa* mites were found in a beehive in Nelson (a small city at the top of the South Island). The government decided not to attempt an eradication, and the mite then began to spread southwards. In 2007 the Varroa Agency began the process of disestablishment,[24] although it took until 2012 for the national pest management plan legislation to be revoked. The economic effects of *Varroa* are difficult to measure, but one estimate for beekeepers was a cost of NZ$12.5 million for 2014 alone.[25]

The incursions of pests such as *Varroa* and fruit fly often come along a specific pathway that also provides a route of entry for other pests. The Biosecurity Act 1993 allows specific pathways to be identified and managed via national pathway management plans. An example could be a plan for the prevention or management of the spread of harmful marine organisms via boat hull fouling. No national pathway management plans currently exist, although both Southland and Northland regional councils have regional pathway management plans for these.[26]

## Regional pest and pathway management plans

The Biosecurity Act 1993 enables the government to determine an overall National Policy Direction for Pest Management, which the 16 regional councils in New Zealand must implement. The last National Policy Direction was initiated in 2015.[27] Each regional council must determine a list of key pests and what should be done about them. Pest species have different thermal or environmental constraints. Most insects, for example, are limited in their distribution by temperature and humidity. These constraints mean that different regions often suffer different pest species. The pests may be a specific problem for a region or be perceived as a high priority after regional councils consult with affected ministers, local authorities, iwi authorities and tribal rūnanga (Māori councils, tribal councils, assemblies, boards or boardrooms). Councils also consider other matters when defining their list of pests, including that the benefits of a control

Two adult female *Varroa* mites, one clinging to the abdomen of this adult bee. *Phil Lester photo*

programme must outweigh the costs.

All the pest species listed in each regional pest management plan are placed into one of four management programmes (Fig. 3.4):

- Exclusion and surveillance: to prevent the establishment of a pest that is present in New Zealand but not yet established in a region.
- Eradication: to reduce a pest species to zero within the region in the short to medium term, typically over the lifetime of the plan (which could be over a period of decades)
- Progressive containment: to contain or reduce the geographic distribution of the pest over time
- Sustained control: to provide for ongoing control of the pest in order to

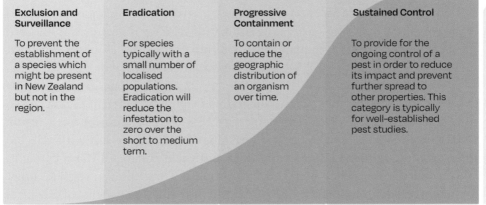

Fig. 3.4: The Biosecurity Act 1993 enables the government to determine an overall National Policy Direction for Pest Management, which the 16 regional councils must implement. The management of each pest named in a regional pest management plan must fall into one of these four categories. Within the Greater Wellington Regional Pest Management Plan 2019–2039, for example, four pests are to be managed under exclusion, including wallaby.[28] Two wallaby species (*Macropus rufogriseus* and *M. eugenii*) are assumed not to be present in the region, so the appropriate response is surveillance with a goal to exclude them. Six species are under eradication, including rooks, which are in small local populations that may be possible to cull and eradicate. Still other pests are well established. Two species of tree (*Pinus* spp. and *Macrocarpa* spp.) are under progressive containment, and six pests including social wasps are under sustained control programmes.

reduce its impacts and its spread to other properties.

In addition, site-led programmes aim to exclude, eradicate, reduce or contain pests or unwanted organisms to protect natural biodiversity at specified sites or defined areas. The sites are typically nature reserves or parks targeted for protection or restoration.

These programmes form a continuum of management options based on levels of infestation and importance in a particular region. The Greater Wellington Regional Council (GWRC), for example, currently has 17 plants and 12 animals on its regional pest management plan. (Note that there are pest groups within the 29 taxa listed on the GWRC regional pest management plan. For example, 'wasps' are listed as one pest but are actually composed of four species. Similarly, 'feral deer' are composed of three species. Many other pests are just a single species.) Four are under exclusion, six under eradication, two under progressive

Wilding conifer spread in 1998 (top left), 2004 (top right), 2015 (bottom left), and 2019 after spraying, Mid Dome, Upper Tomogalak catchment, Southland. The wilding pines are of no commercial value. *Richard Bowman / Biosecurity NZ*

containment, six under sustained control, and 11 under site-led programmes.

One example of a species targeted for eradication by the GWRC is woolly nightshade (*Solanum mauritianum*). This invasive, fast-growing plant produces allelopathic chemicals (toxins) that poison the soil. It is dangerous for people to touch and handle. The regional pest management plan enables GWRC staff to survey, find and kill the plant, or direct landowners to destroy any plants on their land. Residents are not allowed to possess woolly nightshade seeds or living plants.

The rook (*Corvus frugilegus*) is another species defined as a pest under this plan and targeted for eradication. These large birds live conspicuously in breeding colonies or rookeries. Because rooks can be easily disturbed, becoming wary and bait-shy, a key component of the eradication programme is to discourage people from attempting to control rooks themselves. Consequently, under the Biosecurity Act 1993, citizens are not allowed to control these birds. It is an offence to shoot at, poison or interfere with a rookery in any way. People are instead instructed to notify the GWRC of the presence of rooks. It is also illegal to keep, release, spread or sell rooks.

At the other end of the management scale are the sustained control programmes. Four wasp species are recognised as pests by the GWRC: two *Vespula* or yellowjacket species and two *Polistes* or paper wasp species. In high abundance, these wasps are a nuisance, but their stings can cause hospitalisation and even death. In New Zealand, they cause significant cost to the apiculture

A rook at Slimbridge Wetland Centre, Gloucestershire, England. *Adrian Pingstone / Wikimedia Commons*

A paper wasp. *Phil Lester photo*

industry and are major predators and competitors in native forests.[29] The GWRC offers only a minimal sustained control programme against wasps. It will report on the time and location of wasp complaints it receives, refer landowners to pest control companies, potentially release biological control agents, and support initiatives into understanding the impact of wasps on human health. Citizens who are notified by the GWRC that they have a wasp nest on their property have a period of 10 days to destroy the nest, or incur an offence penalty under the Act. Other pests such as rabbits (*Oryctolagus cuniculus*) receive a more intense amount of control and funding under a sustained control programme.

In reality, the number of pests on the GWRC's regional pest management plan is a small subset of the pests present in the region. Some regional councils list more and others fewer on their management plans.[30] To be fair, the government is attempting to manage a number of species through other schemes, such as nine species of plants managed by the National Interest Pest Responses (NIPR) programme coordinated by MPI. The National Pest Plant Accord (NPPA) also aids the government and plant producers by preventing the trade, distribution and propagation of highly damaging pest plants (weeds). The NPPA lists 176 different species as harmful organisms, many of which are listed in appendices in regional pest management plans.

Conspicuously absent from many of the plans are marine species, despite the presence of species recognised by the government as marine pests. Undaria (*Undaria pinnatifida*) is one example of a species that could easily be defined as a pest in need of management. This kelp species is native to the cold, temperate coasts of the northwest Pacific Ocean. It has been listed in 100 of the World's Worst Invasive Alien Species because of its effects on native biodiversity and because of economic concerns over the kelp smothering ships' hulls and aquaculture ventures.[31] It was first found in Wellington Harbour in 1987[32] and is listed on the Official New Zealand Pest Register, although simultaneously and in apparent contradiction the government has allowed its cultivation and farming in the region.[33] Some regions outside of Wellington do have Undaria listed on their regional pest management plans, which perhaps only serves to highlight a level of illogical disparity between those regions where this kelp is cultivated and others where it is a recognised pest.

Could these pest management plans and their legislation be improved? One potential future legislative change to enhance pest management would be to standardise pest lists and management actions between regions. Currently, neighbouring regions with similar climates often have wildly different pests on their management plans. Another improvement that I think is needed is an external auditing process for both national and regional pest management plans. Every few years, the national management agencies and regional councils should be asked whether or not they are achieving their goals. The GWRC set the goal of eradicating all rooks from the region and having no active rookeries within 10 years of the commencement of the plan. Is it likely to achieve that goal with the current rook population trends, management actions and level of funding?

Clearly some regional pest management plans are failing to meet objectives and goals. Otago Regional Council (ORC), for example, lists rabbits as a pest under a sustained control programme. Its stated objective is to prevent rabbits from exceeding a highly qualitative density of 'odd rabbits seen'.[34] The ORC can instruct landowners to control rabbit densities, and can issue a written offence notice for non-compliance. In a *Newsroom* investigation in 2021, Phillip Bunn, a third-generation farmer in Central Otago, gave his assessment of the plan's efficacy:

> There are no rabbit hot spots . . . They're simply everywhere . . . There's no cohesive plan and it's a massive problem. Everybody does their own thing. The Otago Regional Council come and tell us that we need to do something, but that's all they do. And they really have no idea. If you go round our neighbours here, there's people that do a lot. There's people that do absolutely nothing. You can clear a space and the rabbits will just move straight back in within a year.[35]

In response, ORC's biosecurity team leader Richard Lord admitted fault, acknowledged that they had never issued an offence notice to any occupier and indicated they would need to be much more active: 'We've got to be a lot more serious in our approach and our enforcement. And so we've got that new plan in place. The rules are easily applied. There's no excuse for the regional council not to start enforcing and undertaking a lot more inspection work.' The situation has been described as a major ecological disaster.

A similar scenario is developing around the control of introduced wallabies in South Canterbury. Government culling through the South Canterbury Wallaby Board used to keep the numbers of these pests low, but when the board's funding was cut in 1989 the populations began to skyrocket. Wallabies can substantially reduce stocking rates when they occur on farmland. It has been estimated that four wallabies will consume about the same amount of grass as one ewe sheep.[36] Their foraging also stunts grass species diversity and recovery. Pest management by the regional councils in the area has not been working. Ossie Brown, a farmer in the region, has spent 50 years battling rabbits but cites wallabies as an even bigger problem. As he told *New Zealand Geographic* in 2021, 'It's the biggest catastrophe I've ever seen in pest destruction.' In 2016 it was estimated that wallabies cost the country NZ$28 million annually, and that figure is expected to balloon to NZ$83 million by 2025.[37] MPI has launched a programme for the nationwide

At least seven species of Australian wallabies were introduced to New Zealand between 1870 and 1920. They are now a growing pest population, especially in South Canterbury. *Lance Molyneaux / Alamy*

Bennett's Wallaby (*Macropus rufogriseus rufogriseus*). *J.J. Harrison / Wikimedia Commons*

eradication of wallabies under the Biosecurity Act 1993.

Another example is the National American Foulbrood Pest Management Plan, which has the goal of eliminating this bacterial pathogen from managed colonies in New Zealand. Their surveillance data suggests, however, that rates of colony infection over the last five years are higher than they were a decade ago.[38] External audits and consultation could be extraordinarily beneficial for refining targets and methods for both regional and national pest management plans.

## The deliberate introduction of new organisms

The arrival of humans to New Zealand saw the introduction of non-native microbial species, including new diseases, in addition to plants and animals. The Polynesian or Pacific rat (*Rattus exulans*, kiore), was likely introduced to both main islands of New Zealand by some of the first human colonists from central East Polynesia around 1280,[39] who also brought kūmara (sweet potato; *Ipomoea batatas*) and taro (*Colocasia esculenta*). Later, several introductions of potatoes from Europe were recorded between 1769 and 1773. Māori developed potato cultivars from the European-introduced varieties, with dramatic effects on their society. By the early 19th century the potato had overtaken bracken fern rhizome as a staple food, Māori had extensive areas of land cultivated for potato production and a significant proportion of their crop was exported to Australia.[40] Crops and animal introductions over the last thousand years have changed the flora and fauna of New Zealand, sometimes with major economic and social benefit, and sometimes with dire biodiversity consequences.

Times have changed since Captain James Cook deliberately allowed invasive rodents ashore from his boat in 1785.[41] The intentional introduction of species not native to or present in New Zealand is now regulated under the HSNO Act 1996. As explained earlier, 'new organisms' are typically defined as those that have not been recorded as present in New Zealand before 29 July 1998. New organisms may also include those that are genetically modified, species that have been eradicated or eliminated, or organisms that have been imported into the country into containment or quarantine facilities. They can include bacteria, viruses, cell lines for research, sperm or oocytes, seeds, and whole plant and animal organisms. New organisms are important to consider in a context of pest management because intentionally introduced species occasionally become pests, or they might carry passenger pathogens or microorganisms that are harmful to humans or other species. An extremely cautious approach, applying the precautionary principle in the extreme, would be to prevent any new organisms from entering the country. Every potential new organism represents a risk, so the only way to confidently shift the risk to zero is to stop all imports altogether. Careful selection and regulation, however, can limit these risks.

The Environmental Protection Authority (EPA) implements regulatory processes for importing, developing, field testing or releasing new organisms in New Zealand, under the HSNO Act 1996. Any application for importing and releasing a new organism into New Zealand must meet a set of minimum standards, specified in section 36 of the HSNO Act 1996, which states that an application will be declined 'if the new organism is likely to:

- cause any significant displacement of any native species within its natural habitat; or
- cause any significant deterioration of natural habitats; or
- cause any significant adverse effects on human health and safety; or
- cause any significant adverse effect to New Zealand's inherent genetic diversity; or
- cause disease, be parasitic, or become a vector for human, animal, or plant disease, unless the purpose of that importation or release is to import or release an organism to cause disease, be a parasite, or a vector for disease'.[2]

An application must include evidence to satisfy these minimum standards. It needs to describe the need for the new organism, taxonomically describe the species, and identify the beneficial and potential adverse effects. Evidence of Māori engagement is typically essential, given their role as tangata whenua (people of the land) and kaitiakitanga (guardians for the sky, the sea and the land), which is guaranteed under the Treaty of Waitangi.

Applications are submitted to the EPA, which then informs the New Zealand public of the application and invites submissions in support and opposition. The EPA produces an assessment of the application. A decision-making committee is formed to consider the application.

As an example of how the act is used for importing and releasing a new organism, I'll use the application to import and release a butterfly as a biocontrol agent for a plant pest. Japanese honeysuckle (*Lonicera japonica*) is an introduced plant pest that is widespread in New Zealand. It was intentionally introduced to New Zealand and offered for sale in 1872, and it now thrives, climbing over and smothering native plants.[42] This plant is recognised as a harmful organism by the GWRC and is on the Official New Zealand Pest Register. The Department of Conservation Te Papa Awawhai recognises that it is still spreading within New Zealand and is likely to be highly detrimental to conservation values. The search for a biological control agent for Japanese honeysuckle began in 2007 in Honshu, Japan, which is the native range of the plant that climatically matches New Zealand.[43] Herbivorous arthropod species were collected, and three invertebrates were identified that appeared to have potential as a biological control for honeysuckle. After further investigation, one moth species was discarded due to a broad feeding range, while the Honshu white admiral butterfly (*Limenitis glorifica*) and a longhorn beetle (*Oberea shirahatai*) were considered to have potential. Host-range testing for the butterfly in the laboratory proved difficult because of the challenges of rearing the butterfly, but the available data suggested that it is relatively host-specific.

In 2013 the GWRC submitted an application to the EPA proposing to introduce the Honshu white admiral butterfly as a biological control agent for Japanese honeysuckle.

There was a total of 12 submissions on the butterfly application.[44] Seven

Japanese honeysuckle. *May Shimizu*

Honshu white admiral butterfly.
*Alpsdake / Wikimedia Commons*

supported the introduction of the butterfly. The Hawke's Bay Regional Council, for example, considered that 'biological control is now our best option to lessen the effects of this aggressive invader in the area of Hawke's Bay'. Individual submitters, such as Rob Morton, also noted the need for the control of this weed: 'Hand clearance, which we practise, is very difficult. A biological control would be wonderful and give native bush restorers in the North of NZ a great boost in their efforts.' The Bay of Plenty Regional Council noted 'the comments that experiments and studies to date show that the potential negative impacts of the introduction of the white admiral butterfly are negligible. Having intimate knowledge of the vigorous process underpinning the introduction of biological control agents, [we] believe the benefits far outweigh the risks. These are supported by host-range tests, field observations, and literature records.'

This comment – that the introduction was 'supported by host-range tests' – was contrary to the conclusion of at least one individual who opposed the release, Dr Cliff Mason: 'The host testing programme is inadequate, especially as it includes no native species . . . There is no relevant information on the effects of *L. glorifica* herbivory on *L. japonica* . . . The intentional introduction of an alien species is an act of such biological magnitude that it requires very strong justification in the form of a high degree of certainty of beneficial effects and very low probability of adverse effects.' Indeed, the submission from Te Rūnanga o Ngāi Tahu (Ngāi Tahu are the principal Māori iwi of the South Island) noted that the data on host specificity included in the application 'are not very robust', something also noted in the applicant's peer review. Nevertheless, Ngāi Tahu supported the application.

The lack of extensive host-range testing was also considered a problem by Kiwifruit Vine Health, who submitted on behalf of the kiwifruit industry. 'It may be that this biological control agent will have no impact on our industry at all, but as very little effort appears to have been made to understand the risks to our sector it seems inappropriate for the EPA to approve the release of a biological control agent in such situations? . . . I fully appreciate that EPA is required by HSNO to assess the benefits along with the risks but when I look at section 36 of the Act it appears also to be overly focused on risks to native species.'

Decision-making committees have the responsibility of weighing the diverse information, evidence and opinions on issues, including the host range of the biological control agent. In this case, the committee concluded that the butterfly met the minimum standards described above. 'After reviewing the information, the Committee considered any adverse effects, risks or costs from the release of *L. glorifica* to be negligible.'[45] The butterflies were released in 2014 at a site in Waihi in the North Island. 'Vast numbers' were reared in the laboratory, and scientists planned to move progeny throughout the country. The scientists estimated it would take 5–10 years to see a wide-scale impact.[46]

A review published in 2021 found that since the implementation of the HSNO Act 1996, there have been 46 applications for biological control agents to be imported into New Zealand for the control of 30 target species (insect pests and weeds). Of these, nine applications have been for biocontrol of insect pests – eight of these have been approved for release. The one species that was not approved was a generalist biological control agent with a broad host diet. The authors of the review lamented that, both nationally and internationally, the importation and release of generalist species with a broad diet is becoming increasingly unlikely.[47]

The potential risks and level of precaution applied to the importation and release of a new organism can be very contentious, emphasised by the recent hotly debated introduction of dung beetles. It has been estimated that cows in New Zealand produce 85 million tonnes of dung every year. Because New Zealand evolved without large native mammals, there are no native dung beetles to utilise this resource. The application to the EPA to introduce 11 species of dung beetles as 'new organisms' in 2010 was hotly debated. Proponents thought the beetles would

Dung beetles (*Geotrupes* species) in fresh cow dung in a pasture. *Richard Becker / Alamy*

reduce the amount of dung sitting on the soil surface and running into rivers after rainfall; buried dung could improve soil quality and reduce the abundance of flies and internal parasites of livestock.[48] Critics thought the introduction could increase rat populations or compete with native beetles.[49] Dr Denise Barnfather, medical officer of health with the Regional Public Health Service, suggested the beetles might spread nasty gut diseases for humans and concluded: 'Until the significance of these potential human exposures to pathogens harboured by dung beetles are adequately researched and understood, public health recommends that a precautionary approach is taken and that exotic dung beetles are not introduced into New Zealand.'[50] After a committee received written submissions and heard presentations, a decision was made to allow the beetle release in 2011. The principal scientist for the Environmental Protection Authority said, 'It is not unusual for different experts to come to different conclusions about decisions on the release of new plants and animals, however the application went through a robust process, and the EPA stands by the decision.'

## Importing new species on a regular basis: 'Import Health Standards'

Early Māori brought kūmara to New Zealand; later, Pākehā settlers brought potatoes. But imagine you are seeking to bring a new plant into the country today. What would you have to do?

In order to import nursery stock (such as plants, tissue cultures and bulbs) or seeds, you would need a determination of whether the species is a new organism under the HSNO Act 1996. MPI maintains a Plants Biosecurity Index (PBI), a register of nearly 30,000 approved plant species.[51] This database identifies whether the plant is present in New Zealand, the requirements for importing seed for sowing and nursery stock, and whether they can be released. Both potatoes and kūmara are listed on the database. If your plant was not on the database, you'd need to make an application to the EPA, similar to that described for biological control agents above, to import the species as a new organism.

The next thing you'd need is an import health standard (IHS), required to import any biosecurity risk goods into New Zealand. Risk goods include anything that could constitute, harbour or contain an organism that may cause unwanted harm to natural or physical resources, or to human health in New Zealand. An IHS is required for plants and plant products, animals and animal products, food, and even imported shipping containers and vehicles. The major component of an IHS is an assessment of the risk, for which MPI uses the definition from the SPS Agreement: 'the evaluation of the likelihood of entry, establishment or spread of a pest or disease . . . and of the associated potential biological and economic consequences'.[52] The SPS Agreement is the foundation for an evidence-based analysis that produces a qualitative assessment needed for an IHS. The 'import risk analysis' considers the likelihood that the goods will import organisms, their potential impact on human health and the environment, and what could be done to manage the risk. IHSs are administered under the Biosecurity Act 1993, which requires that any standard has industry consultation while being developed.

The government database of IHSs shows an astounding diversity of goods that are imported.[53] You can find them for alligator meat imported from the United States, durian fruit from Thailand and frozen Nile perch from Kenya.

Perhaps the key question when importing any living species is what else do these imports bring with them as passengers? Passenger species are nearly always a critical issue with the intentional introduction of any species. One of the most recent examples of a harmful pathogen that was unintentionally introduced with an intentionally imported organism is the bacterial species *Pseudomonas syringae* pv. *actinidiae* that has devastated the kiwifruit industry in New Zealand. As discussed earlier, Psa is one of the three species under a national pest management plan. Its introduction is hypothesised to have come in a consignment of kiwifruit pollen that was legally imported into New Zealand from China, for the new use of commercial artificial pollination of kiwifruit orchards. Pollen has been known to play a role in spreading diseases such as Psa for over 50 years. A group of 212 kiwifruit orchardists brought a class-action suit alleging that the government was liable for losses caused by allowing the pollen imports that likely contained the devastating plant disease. In 2021 MPI agreed to a settlement with the industry for NZ$40 million. In a report following the settlement, Kiwifruit Claim chair John Cameron said it concluded what had been 'a very long, hard and stressful fight':

> What happened to the kiwifruit industry in 2010 was entirely preventable – MPI knew Psa existed, they knew the damage it could cause if it was let into the country, and they breached their own protocols that were in place to keep it out. It is critical our primary industries can rely on our government to perform their biosecurity role with reasonable care and skill; we don't want to ever see this happen again.[20]

Similar concerns were raised over the potential of passenger pathogens in honey that could affect honey bees. This related to a proposed IHS developed by the government. After considerable debate through the media and courts, the international importing of honey into New Zealand has been almost entirely stopped (See 'Case study 3: Honey bee disease and allowing honey imports into New Zealand', page 115). This case highlights the difficulties in balancing risk, trade agreements, and potential or perceived protectionism by industries.

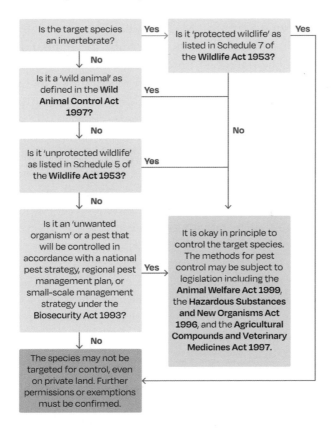

Fig. 3.5: A flow chart for determining legislative approval for pest control. A range of legislative acts determine whether it is lawful to control a particular pest. That there are exceptions, such as with the control of rooks under a Regional Pest Management Plan. Rooks are classified as a pest for eradication by regional councils, but typically rook control is tightly regulated: it must be carried out by councils or under council supervision. *Developed from National Pest Control Agencies (2018)*[54]

## Summary

I'm not envious of the roles that biosecurity officials and pest managers play. Both are faced with limited amounts of money and a diverse array of public opinions on how this funding should be spent. The question of which pests or pathways should be a priority is typically answered through the lens of each individual's industry commitments or preferences for leisure activities. In addition, the publicity that both biosecurity officials and pest managers receive is overwhelming when systems fail. MPI makes the evening news when fruit flies are discovered to have breached the border, while the thousands of times that pests are stopped at the border go unreported. Similarly, there are likely to be many pests that are successfully managed by regional pest management plans. But those that aren't, such as rabbits and wallabies in the South Island, are the species that receive media attention.

The general public and scientists also have many opinions on precaution. As discussed in Chapter 2, a precautionary approach is often applied when extensive scientific knowledge is lacking. It emphasises caution to pause and review before accepting and implementing new innovations, products or services that may prove to have deleterious consequences. Acceptable levels of precaution, however, differ substantially between individuals. You and I will differ in the level of scientific knowledge we need to satisfy the minimum standards for the release of a new organism, or to evaluate the risks of importing a commodity like honey from Australia. An optimal application of legislation is difficult, and perhaps the norm is that not everyone will be satisfied. Any regulatory regime involves trade-offs, especially in trade negotiations, that the public are often unhappy with. Those who are most unhappy often shout the loudest and get the media attention.

The legislation regarding the introduction of new organisms, biosecurity and pest management can be enhanced. The government has announced that in the coming years it will review the Biosecurity Act 1993. I hope they will consider how national and regional pest management plans could be made more effective. Biosecurity and new organisms in the marine environment also need much more emphasis in the future. The review of the act will also need to incorporate a major ruling from New Zealand's Waitangi Tribunal. (The Waitangi Tribunal, Te Rōpū Whakamana i te Tiriti o Waitangi, is a permanent commission of inquiry established under the Treaty of Waitangi Act 1975. It investigates and makes recommendations on claims brought by Māori relating to actions and omissions of the Crown that have breached promises made in the Treaty of Waitangi, signed in 1840 by representatives of the British Crown and Māori chiefs or rangatira.) The Wai 262 claim, also known as the 'flora and fauna claim', has importation ramifications for biodiversity management. In response to this claim, the New Zealand government proposed ministerial oversight groups for three broad kete (baskets) of issues that include the genetic and biological resources of taonga (treasured) species, the relationship with the environment, and the conservation estate. The role of Māori as kaitiaki (guardians) will need to be included in legislative changes in recognition of partnership between the Crown and Māori.

## Case study 1: Prions, mad cows and sheep, and cannibalism

Prions are an unusual entry on New Zealand's list of 'unwanted organisms'. Prions are misfolded proteins: they aren't actually organisms, but are infectious, as they transmit their misfolded shape onto normal variants of the same protein. Prion diseases are characterised by several fatal and transmissible neurodegenerative diseases in humans and other animals, including cows and sheep.

One of the most famous cases of prions in humans is from Kuru disease.[55] This incurable and fatal neurodegenerative disorder was common among the Fore people of Papua New Guinea, especially among women and children. The name 'Kuru' is derived from the Fore word *kuria* or *guria*, referring to body tremors or shaking, which are classic symptoms. Sufferers also exhibited pathologic bursts of laughter, so the disease also became known as 'laughing sickness' or the 'smiling death syndrome'.

The cause of Kuru was initially unknown, and some attributed it to sorcery or witchcraft by the Fore people. An American physician and a regional medical officer hypothesised that cannibalism was a possible cause, due to a mortuary custom: the Fore people believed that eating the body of the dead helped to free the spirit. The entire body had to be eaten, including the brain. Children and women would consume the brain, but boys over the age of six were excluded from the ritual.[56]

After injecting infected brain material into chimpanzees, the American team proved that an unknown disease factor caused the disease. When cannibalism ended in the 1950s and 60s, the epidemic declined sharply in the Kuru people.

Prions cause disease in animals and humans. Examples include chronic wasting disease (CWD) in deer, bovine spongiform encephalopathy (BSE) in cattle (commonly known as mad cow disease), and Creutzfeldt–Jakob disease (CJD) in humans. The last confirmed incursion and infection of prions in New Zealand was in sheep on Mana Island in 1978. A flock of east Friesian and Finn sheep had been imported from the UK to quarantine on this island. Initially, just one ewe was observed with a prion disease called scrapie. All the closely related sheep from the same flock were killed and burnt. Later that year, however, the disease reappeared. The entire flock of 1900 sheep had to be slaughtered, burnt and buried on the island.

Scrapie is a well recognised problem in the UK and Europe, which was the source of New Zealand's sheep. The disease was observed in sheep that were imported in the 1950s, which were killed, and the disease eradicated. The system of quarantining sheep on isolated islands such as Mana is a key reason why New Zealand is now scrapie-free, although some degree of luck in breeds from areas in the UK played a role too.[57]

Sheep imported from the UK in 1978 were found to be infected with scrapie. Mana Island was used as a quarantine island, and the slaughtered sheep had to be burned so that the disease did not spread to mainland sheep. *Ken Seccombe*

## *Case study 2: 55,000 fruit trees might be diseased and have to be destroyed*

Imports may have passenger diseases or pests not currently present in New Zealand. Section 22 of the Biosecurity Act 1993 enables the development of Import Health Standards, which are effectively a restriction on international trade. Their implementation is subject to international law under the WTO Agreement on the Application of the Sanitary and Phytosanitary Measures. New Zealand sets its own 'appropriate level of protection' under the SPS Agreement. These standards prescribe the requirements for 'risk goods' that must be met before the goods can be imported.

Apple and stone-fruit growers in New Zealand import new plant stock material from around the world. A new cherry cultivar might be developed in the United States which could be of significant value, so growers will seek to purchase and import live tree cuttings. These cuttings are classified as risk goods. The applicable Import Health Standards are called 'MPI Standard 155.02.06: Importation of Nursery Stock'. Foreign countries and companies exporting cuttings must comply with this standard and are audited periodically.

One such facility exporting trees to New Zealand was the Clean Plant Center

Grafted cuttings on a cherry tree in an orchard in Luberon, France.
*BIOSPOHOTO / Alamy*

Northwest – Fruit Trees, located in Washington. An audit in 2018 identified that many testing records were missing, and in some cases test results suggested that the material had failed the relevant tests. The USDA also conducted an audit of the Clean Plant Center and concluded that there had been a 'systemic breakdown in record-keeping and adherence to New Zealand's requirements for maintaining approval as an off-shore plant quarantine facility'.[58]

The Ministry for Primary Industries ordered the destruction of 55,000 apple and stone-fruit cuttings that had been imported and grown into trees over 2012–17. Ministry response manager Dr John Brightwell said there was no choice but to take a precautionary approach and to act fast: 'These actions demonstrate how seriously we take biosecurity and the high expectations we place on the assurances given by our overseas trading partners.'[59] MPI concluded that the goods were unauthorised and it was illegal to hold them. There was no evidence of pests and diseases on the tree cuttings, but because of the missing or incomplete documentation, MPI had no confidence that the material met the Import Health Standard.

Several nurseries and orchardists together challenged the seizure and the ministry's directions in the High Court in 2018. The High Court ruled in favour of the nurseries and orchardists. It concluded that failure of the Clean Plant Center's record-keeping was insufficient to justify the destruction of all trees imported over a six-year period, especially when there was no evidence of pests or diseases on these trees. Further, the High Court found that most of the seized fruit trees were no longer 'goods' under the act. The trees derived from the imports were now outside the scope of the section of the act used to 'seize and dispose of unauthorised goods' (Section 116).[1] By this time it was too late for many growers, who had complied with the orders and destroyed their trees.

The response from MPI was to use alternative legislation, on the 'Power to give directions' (Section 122), to direct the growers to treat goods that 'may be contaminated with pests or unwanted organisms; or to destroy . . . any organism or organic material or thing that there are reasonable grounds to believe harbours a pest or unwanted organism'.[1]

## Case study 3: Honey bee disease and allowing honey imports into New Zealand

Honey bees are vital for pollination in the agricultural industries, and they produce honey for domestic and international markets. Minimising disease in bees is vital. New Zealand has many honey bee pathogens, but there are several that are not present here, including the bacteria that cause European foulbrood disease (*Melissococcus plutonius*). The OIE World Organisation for Animal Health lists European foulbrood as an important disease in the Terrestrial Animal Health Code, which obligates countries to report the presence of the bacteria and for importing countries to implement measures to stop its introduction.[60]

Beekeepers in Australia have European foulbrood but have long wanted market access for their honey to New Zealand, just as beekeepers here have access to Australia. Scientists from the government agency Biosecurity NZ previously recognised that European foulbrood is likely to be present in honey. A report concluded that the risk of its release into the country would be 'non-negligible'. But it also asked 'how cautious decision makers should be in the face of imperfect scientific information', and went on to say:

> Scientific uncertainty is a fact of life, and unanimous agreement among stakeholders and scientists is rarely possible. 'Zero risk' is not attainable, whether trade occurs or not, and the acceptability of any particular import risk depends on stakeholder perceptions as to the distribution of the benefits from the trade in question. Thus, acceptable risk decisions are essentially political judgements that attempt to reflect societal values, and such decisions can at best be informed by science without being purely scientific.[61]

Consequently, in 2006 Biosecurity NZ issued a new Import Health Standard for honey and related bee products from Australia. It concluded that 'honey could be imported from countries where European foulbrood is present, provided it was subject to heat treatment giving a million-fold reduction in bacteria. This means that 99.9999% of the bacteria will have been killed.'[62]

New Zealand beekeepers were furious at this decision and feared that lifting a ban on Australian honey imports would result in the end of their industry. The then agriculture minister, David Carter, responded that the bee industry was 'overly pessimistic' and suggested there was a degree of protectionism at work.

Honey bees on a frame used for queen rearing.
*Phil Lester photo*

He argued from a position of fair trade: 'We are an export nation and we want fair access to global markets and, in this case, Australia is saying we want fair access for honey and we say we are investigating the biosecurity risks.'[63]

The Beekeepers Association of New Zealand took the case to the High Court, which validated and supported Biosecurity NZ's decision to allow imports. However, the High Court also noted that honey should be subject to strict regulations from the Environmental Risk Management Authority (ERMA; now the Environmental Protection Authority or EPA) regulations before it was cleared. ERMA operated according to the Hazardous Substances and New Organisms Act 1996. This act states that an organism such as *Melissococcus plutonius*, that is not recorded as present in New Zealand prior to July 1998, would be a 'new organism'. A person intending to import a new organism must apply for approval. Even potential passenger organisms, such as European foulbrood disease, would be 'new organisms'. This ruling had major implications for all imports that have potential passenger species associated with them, not just honey from Australia.

## Further reading and discussion

1. Only a single application for a exotic generalist arthropod biological control agent has been made since the implementation of the Hazardous Substances and

New Organisms (HSNO) Act 1996. This application, to introduce a potential biological control agent with a broad dietary range, was denied. Yet, these generalist species can be of huge benefit. What information and analysis would be needed to enable their introduction in the future?

*See:* Gerard, P. J., & Barratt, B. I. P. (2021). Risk assessment procedures for biological control agents in New Zealand: Two case studies for generalists. *BioControl, 66*, 143–150. doi.org/10.1007/s10526-020-10049-4

2. What issues should be considered when developing a public process for decision-making on the introduction of a new organism as a biological control agent?

*See:* Wilkinson, R. & Fitzgerald, G. (1997). Public perceptions of biological control of rabbits in New Zealand: Some ethical and practical issues. *Agriculture and Human Values, 14*, 273–282. doi.org/10.1023/A:1007473215360. This paper discusses the public perception of risk and other topics using the examples of rabbits as a key vertebrate pest.

3. Kauri dieback in New Zealand is caused by *Phytophthora agathidicida,* a microscopic fungus-like organism. In 2021 the government announced it would

Kauri dieback, caused by *Phytophthora agathidicida*, can kill kauri of all ages. *MPI*

be managed under a National Pest Management Plan. Is this the right approach? What is needed to achieve effective management? How would your approach to the management of kauri dieback differ from what has been proposed?

*See:* Bradshaw, R. E., et al. (2020). *Phytophthora agathidicida*: Research progress, cultural perspectives and knowledge gaps in the control and management of kauri dieback in New Zealand. *Plant Pathology 69*(1), 3–16. doi. org/10.1111/ppa.13104

4. The pests listed on regional pest management plans differ substantially between regions within New Zealand. How much of a problem are these differences in pest priorities, funding and resources? What is the solution?

*See:* Brenton-Rule, E. C., Frankel, S., & Lester, P. J. (2016). Improving management of invasive species: New Zealand's approach to pre- and post-border pests. *Policy Quarterly 12*(1), 17–25. doi.org/10.26686/pq.v12i1.4582

GREATER WELLINGTON REGIONAL
**PEST MANAGEMENT PLAN**
2019–2039

greater WELLINGTON
REGIONAL COUNCIL
Te Pane Matua Taiao

The Greater Wellington Regional Pest Management Plan 2019–2039 was launched in 2019. Pictured is a titipounamu (rifleman) during translocation from Wainuiomata to Zealandia. *Chris Gee / GWR Te Pane Matua Taiao*

# References

1. Biosecurity Act 1993. legislation.govt.nz/act/public/1993/0095/latest/whole.html

2. Hazardous Substances and New Organisms Act 1996. legislation.govt.nz/act/public/1996/0030/latest/DLM381222.html

3. Wildlife Act 1953. legislation.govt.nz/act/public/1953/0031/latest/whole.html

4. Wild Animal Control Act 1977. legislation.govt.nz/act/public/1977/0111/latest/whole.html

5. Agricultural Compounds and Veterinary Medicines Act 1997. legislation.govt.nz/act/public/1997/0087/latest/whole.html

6. Ministry for Primary Industries. (2021, August). *A biosecurity team of 4.7 million*. mpi.govt.nz/biosecurity/about-biosecurity-in-new-zealand/biosecurity-2025/biosecurity-2025/a-biosecurity-team-of-4-7-million/

7. Biosecurity New Zealand (2020, November 10). *Official New Zealand pest register* [incorporating the former *Unwanted organisms register*]. Ministry for Primary Industries. pierpestregister.mpi.govt.nz/PestsRegister/ImportCommodity/

8. Biosecurity (Notifiable Organisms) Order 2016. legislation.govt.nz/regulation/public/2016/0073/9.0/whole.html

9. MacLellan, R., King, K., McCarthy, B., et al. (2020, September). National Fruit Fly Surveillance Programme annual report. *Surveillance, 47*(3), 98–101. Ministry for Primary Industries. sciquest.org.nz/node/165130

10. Ministry for Primary Industries. (2016, November 21). Fruit fly stopped at the border [Media release]. mpi.govt.nz/news/media-releases/fruit-fly-stopped-at-the-border/

11. Daniels, C. (2020, January 30). Auckland's $18 million Queensland fruit fly hunt. *NZ Herald*.

12. Fruit Fly Council. (2017). *New Zealand Fruit Fly Strategy 2017–22*. Government Industry Agreement for Biosecurity Readiness and Response. gia.org.nz/Portals/79/Content/Documents/Resource-Library/NZ%20Fruit%20Fly%20Strategy%202017-22.pdf?ver=2017-07-28-110855-373

13. Acosta, H., Earl, L., Growcott, A., et al. (2020, January). *Atlas of biosecurity surveillance*. Surveillance and Incursion Investigation Group, Biosecurity New Zealand, Ministry for Primary Industries. mpi.govt.nz/dmsdocument/39674/direct

14. Christy, M. T., Yackel Adams, A. A., Rodda, G. H., et al. (2010). Modelling detection probabilities to evaluate management and control tools for an invasive species. *Journal of Applied Ecology, 47*(1), 106–113. doi.org/10.1111/j.1365-2664.2009.01753.x

15. Tyrrell, C. L., Christy, M. T., Rodda, G. H., et al. (2009). Evaluation of trap capture in a geographically closed population of brown treesnakes on Guam. *Journal of Applied*

*Ecology, 46*(1), 128–135. doi.org/10.1111/j.1365-2664.2008.01591.x

16. Seymour, A., Varnham, K., Roy, S., et al. (2005). Mechanisms underlying the failure of an attempt to eradicate the invasive Asian musk shrew *Suncus murinus* from an island nature reserve. *Biological Conservation, 125*(1), 23–35. doi.org/10.1016/j.biocon.2005.03.005

17. Biosecurity (National Bovine Tuberculosis Pest Management Plan) Order 1998. legislation.govt.nz/regulation/public/1998/0179/latest/whole.html

18. Biosecurity (National American Foulbrood Pest Management Plan) Order 1998. legislation.govt.nz/regulation/public/1998/0260/latest/whole.html

19. Biosecurity (National Psa-V Pest Management Plan) Order 2013. legislation.govt.nz/regulation/public/2013/0139/10.0/whole.html

20. Morrison, T. (2021, February 13). Government pays $40m to settle long-running kiwifruit Psa claim. *Stuff.* stuff.co.nz/business/300229318/government-pays-40m-to-settle-longrunning-kiwifruit-psa-claim

21. Vanneste, J. L. (2017). The scientific, economic, and social impacts of the New Zealand outbreak of bacterial canker of kiwifruit (*Pseudomonas syringae* pv. *actinidiae*). *Annual Review of Phytopathology, 55*, 377–399. doi.org/10.1146/annurev-phyto-080516-035530

22. Lester, P. J. (2020). *Healthy bee, sick bee: The influence of parasites, pathogens, predators and pesticides on honey bees*. Victoria University Press.

23. Biosecurity (National (South Island) Varroa Pest Management Strategy) Order 2005. legislation.govt.nz/regulation/public/2005/0017/latest/whole.html

24. Varroa Agency Inc. (2007, September 26). Disestablishment of the Varroa Agency Inc. [Press release.] Scoop Media. scoop.co.nz/stories/BU0709/S00518.htm

25. Iwasaki, J. M., Barratt, B. I. P., Lord, J. M., et al. (2015). The New Zealand experience of varroa invasion highlights research opportunities for Australia. *Ambio, 44*, 694–704. doi.org/10.1007/s13280-015-0679-z

26. Northland Regional Council. (2018). *Northland Regional Pest and Marine Pathway Management Plan 2017–2027.* consult-nrc.objective.com/portal/biosecurity/rpmp

27. National Policy Direction for Pest Management 2015. mpi.govt.nz/dmsdocument/9464-National-Policy-Direction-for-Pest-Management-2015

28. Greater Wellington Regional Council. (2019). *Greater Wellington regional pest management plan 2019–2039.* gw.govt.nz/document/16831/greater-wellington-regional-pest-management-plan-2019-2039

29. Lester, P. J. (2018). *The vulgar wasp: The story of a ruthless invader and ingenious predator.* Victoria University Press.

30. Brenton-Rule, E. C., Frankel, S., & Lester, P. J. (2016). Improving management of invasive species: New Zealand's approach to pre- and post-border pests. *Policy Quarterly, 12*(1), 17–25. https://doi.org/10.26686/pq.v12i1.4582

31. Lowe, S., Browne, M., Boudjelas, S., et al. (2000). *100 of the world's worst invasive alien species: A selection from the global invasive species database.* Invasive Species Specialist Group. issg.org/pdf/publications/worst_100/english_100_worst.pdf

32. Hay, C. H., & Luckens, P. A. (1987). The Asian kelp *Undaria pinnatifida* (Phaeophyta: Laminariales) found in a New Zealand harbour. *New Zealand Journal of Botany, 25*(2), 329–332. doi.org/10.1080/0028825x.1987.10410079

33. Ministry of Agriculture and Forestry. (2012, January 19). Areas designated for Undaria farming. [Press release.] *Scoop.* scoop.co.nz/stories/AK1201/S00331/areas-designated-for-undaria-farming.htm

34. Otago Regional Council. (2019). *Otago pest management plan 2019–2029.* orc.govt.nz/media/10068/orc-regional-pest-management-plan-2019-29_final_corrected-21.pdf

35. Reid, M. (2021, May 4). Rabbits: 'It's as bad as it's ever been'. *Newsroom.* newsroom.co.nz/rabbits-its-as-bad-as-its-ever-been

36. Latham, A. D. M., Latham, M. C., Norbury, G. L., et al. (2019). A review of the damage caused by invasive wild mammalian herbivores to primary production in New Zealand. *New Zealand Journal of Zoology, 47*(1), 20–52. doi.org/10.1080/03014223.2019.1689147

37. Morris, B. (2021, May–June). The occupation. *New Zealand Geographic, 169.* nzgeo.com/stories/the-occupation/

38. The Management Agency National American Foulbrood Pest Management Plan. (2020). *Annual Report: 2019/2020.* afb.org.nz/wp-content/uploads/2020/12/2019-20-Annual-Report.pdf

39. Wilmshurst, J. M., Anderson, A. J., Higham, T. F. G., et al. (2008). Dating the late prehistoric dispersal of Polynesians to New Zealand using the commensal Pacific rat. *Proceedings of the National Academy of Sciences, 105*(22), 7676–7680. doi.org/10.1073/pnas.0801507105

40. Harris, G. (2001). Conservation of relict potato *Solanum tuberosum* cultivars within Māori communities in New Zealand. *Pacific Conservation Biology, 7*(3), 204–213. doi.org/10.1071/pc010204

41. Atkinson, U. A. E. (1973). Spread of the ship rat (*Rattus r. rattus* L.) in New Zealand. *Journal of the Royal Society of New Zealand, 3*(3), 457–472. doi.org/10.1080/03036758.1973.10421869

42. Williams, P. A., & Timmins, S. M. (1998). Biology and ecology of Japanese honeysuckle (*Lonicera japonica*) and its impacts in New Zealand. *Science for Conservation, 99.* Department of Conservation. doc.govt.nz/globalassets/documents/science-and-technical/sfc099.pdf

43. Paynter, Q., Konuma, A., Dodd, S. L., et al. (2017). Prospects for biological control of *Lonicera japonica* (Caprifoliaceae) in New Zealand. *Biological Control, 105*, 56–65. doi.org/10.1016/j.biocontrol.2016.11.006

44. Environmental Protection Authority. (2013). *APP201710: Submissions.* epa.govt.nz/assets/FileAPI/hsno-ar/APP201710/08ef96f1cf/APP201710-APP201710-Submission-Document.pdf

45. Environmental Protection Authority. (2013). *APP201710: Decision.* epa.govt.nz/assets/FileAPI/hsno-ar/APP201710/b3ff48bd89/APP201710-APP201710-Decision-Document-Amended-V1.pdf

46. *Katikati Advertiser.* (2018, February 15). Butterflies bred to fight Japanese honeysuckle pest.

47. Gerard, P. J., & Barratt, B. I. P. (2021). Risk assessment procedures for biological control agents in New Zealand: Two case studies for generalists. *BioControl, 66,* 143–150. doi.org/10.1007/s10526-020-10049-4

48. Mandow, N. (2020, January 31). Beetle mania. RNZ. rnz.co.nz/programmes/two-cents-worth/story/2018730738/beetle-mania

49. Beggs, J. (2013, March 19). Dung beetles pose threat to us and our wildlife. *NZ Herald.*

50. Stone, A. (2013, March 9). Fears new beetles will spread diseases. *NZ Herald.*

51. Ministry for Primary Industries. (n.d.). *Plants Biosecurity Index (Version: 02.01.00).* Retrieved May 10, 2021 from maf.govt.nz/cgi-bin/bioindex/bioindex.pl

52. Ministry for Primary Industries. (2020). *Import risk analysis: Process overview.* mpi.govt.nz/dmsdocument/41779-Import-risk-analysis-process-overview

53. Ministry for Primary Industries. (2020). *Import health standards (IHS).* mpi.govt.nz/legal/compliance-requirements/ihs-import-health-standards/

54. National Pest Control Agencies. (2018). Legislation guide: User guide to legislation relating to terrestrial pest control, 35.

55. Liberski, P. P. (2013). Kuru: A journey back in time from Papua New Guinea to the Neanderthals' extinction. *Pathogens, 2*(3), 472–505. doi.org/10.3390/pathogens2030472

56. Whitfield, J. T., Pako, W. H., Collinge, J. & Alpers, M. P. (2008). Mortuary rites of the South Fore and kuru. *Philosophical Transactions of the Royal Society B: Biological Sciences, 363* (1510), 3721–4. doi.org/10.1098/rstb.2008.0074

57. Bruère, A. N. (2003). Scrapie freedom – the New Zealand story. *Surveillance, 30*(4), 3–7.

58. Waimea Nurseries Limited V Director-General for Primary Industries (2018, August 23). NZHC 2183.

59. Frykberg, E. (2018, June 7). No biosecurity concern over 55,000 seized cuttings – industry. RNZ. rnz.co.nz/news/country/359050/no-biosecurity-concern-over-55-000-seized-cuttings-industry

60. OIE World Organisation for Animal Health. (2021). Diseases of bees. oie.int/en/disease/diseases-of-bees/

61. Pharo, H. J. (2006). Risk of European foulbrood in imported honey bee products.

*ISVEE 11: Proceedings of the 11th Symposium of the International Society for Veterinary Epidemiology and Economics,* 615–627.

62. Biosecurity NZ. (2006, July 11). Honey Import Health Standard issued. *Scoop.* scoop.co.nz/stories/SC0607/S00021/honey-import-health-standard-issued.htm

63. Donaldson, M. (2010, August 1). Beekeepers fear lift of import ban. *Sunday Star-Times.*

# 4. ERADICATION

## To kill every last one

One of the very first successful eradications in New Zealand was unplanned. In 1986 conservation authorities decided to try dropping toxic bait from a helicopter onto the 173-hectare Moutohorā Island in the Bay of Plenty for the control of rabbits (*Oryctolagus cuniculus*). The team used a variety of different baits, such as carrots and wax blocks, laced with pesticides including 1080 (sodium fluoroacetate), brodifacoum and bromodiolone. Rabbits were successfully eradicated from the island. But unexpectedly, introduced Norway rats also found the bait and pesticides irresistible and every last one of them died too.[1] The aerial use of bait laced with 1080 or brodifacoum has since been developed as a mainstay for eradication and pest suppression around the globe. This form of pest eradication typically refers to the complete culling or removal of an entire population of a target species from a defined area, such as an island. Rabbits were eradicated from Moutohorā Island, but still hop around in abundance just a few kilometres away on mainland New Zealand.

Microbiologists have quite a different definition of eradication. They think of it more on a global scale, as in Thomas Aiden Cockburn's definition of 'the extinction of the pathogen that causes disease'.[2] Consequently, for many microbiologists, even the viruses that cause the devastating diseases of smallpox and rinderpest are not eradicated. Both viruses are no longer seen in human or wildlife populations, but they are held in (hopefully very secure) laboratories. Another definition of eradication in microbiology that is widely used and accepted by many organisations, including the WHO, is the 'permanent reduction to zero of the worldwide incidence of infection caused by a special agent as a result of deliberate efforts'.[2] The last known smallpox case was in Somalia in 1977, and the last reported case of rinderpest occurred in Kenya in 2001. So, under the WHO definition, both smallpox and rinderpest viruses would be considered eradicated.

Instead of the word 'eradication', many microbiologists use the term 'elimination' for country- or area-specific disease control. Angel Corona explains this distinction in his article on disease eradication: 'Disease elimination does not require the *permanent worldwide* reduction of disease incidence to zero, but rather reducing incidence to zero in a particular geographic area. One example of this difference would be the elimination of cholera from countries like Peru, despite *Vibrio cholera* not being eradicated globally.'[2] As I write this in 2021, New Zealand is attempting to eliminate the bacterial cattle pathogen *Mycoplasma bovis* (Fig. 4.1). *Mycoplasma bovis* can cause range of serious conditions in cattle, including mastitis that does not respond to treatment, pneumonia, arthritis and

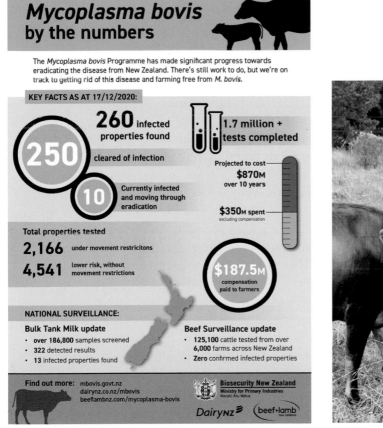

Fig. 4.1: As of 2022, *Mycoplasma bovis* is under an eradication programme in New Zealand. *Ministry for Primary Industries and licensed by MPI for reuse under the Creative Commons Attribution 4.0 International licence / Phil Lester photo*

late-term abortions. The bacteria is classified as an unwanted organism under the Biosecurity Act 1993. The eradication programme relies on movement restrictions and the culling of infected herds. An estimated 126,000 cattle will need to be killed and the programme will likely cost near a billion dollars over a 10-year period. Many microbiologists would refer to this as a disease 'elimination' programme, rather than eradication, because it refers to national rather than global disease management.

In this chapter I'll focus on eradication from a conservation and biodiversity perspective, returning to pathogen and disease elimination or eradication later. The chapter will start with a discussion of eradication scale, and the history of successful and failed pest eradications. Which pests should be considered for eradication? When should we attempt eradication, or decide instead to use ongoing pest control? How should biodiversity managers decide?

## An increasing scale of pest eradication programmes

At the smallest scale, the country-specific eradication of newly arrived pests happens frequently. Quarantine officials will find and kill an insect, or a plant seed, in the suitcase of a passenger returning from an overseas holiday. An inspector will find spiders in a container of furniture unloaded from a ship that has travelled halfway around the world. The container will be fumigated with pesticides such as methyl bromide, killing animals hiding in every crevice. These are small eradications, of one or just a few individuals that might otherwise have gone on to establish and spread. They don't make the evening news or cost a lot of money.

At the other end of the scale are eradication campaigns that encompass entire islands. These are massive programmes that take years, cost many millions of dollars and can have substantial biodiversity benefits. The largest of these, to date, has been the eradication of rats and mice from South Georgia.

South Georgia is a 165-kilometre-long island lying 1400 kilometres to the east and south of the Falkland Islands and Argentina. It is cold, with a polar climate, and is isolated and inhospitable. The landscape is largely barren with no trees. The mountain slopes are furrowed with deep glaciers that run to the sea. The unique flora and fauna of South Georgia had evolved without rodents. Captain James Cook landed on the island in 1775 and claimed it for Great Britain.

Subsequently, whaling, sealing and fishing stations were established. With boats came Norway rats (*Rattus norvegicus*) and house mice (*Mus musculus*). Many species were affected, including the South Georgia pipit (*Anthus antarcticus*) and a variety of burrow- or cavity-nesting seabirds, which were unable to breed in the presence of rats. Even prions and petrels nesting on high mountain slopes were attacked. After the introduction of rodents, many of these birds survived only on the few surrounding offshore islands.

In the early 2000s a charity called the South Georgia Heritage Trust was formed. They raised funds to eradicate rodents and assumed the management of the operation, which ended up costing GBP7.5 million donated by private individuals and foundations. It was decided that the best approach would be using helicopters to spread the toxin brodifacoum, which had been a successful method for rodent eradication on smaller islands elsewhere. The trust did have concerns. Specifically, the toxin would likely have non-target effects, killing some birds. People stationed on the island could be affected too if the poison contaminated water supplies. Also, there might be nest mortality from aircraft flying over nesting birds, because rats and mice residing in colonies of albatross and penguins would have to be baited.

A helicopter spreading baits laced with brodifacoum for a successful rodent eradication on South Georgia. Some bird by-kill was observed after the helicopter baiting, but populations recovered quickly. *Roland Gockel photo*

Helicopters were needed because the terrain was not suitable for ground baiting, but the island's subantarctic geography did offer advantages. Not all of the island's landmass was infested with rodents. The glaciers that ran from the mountains to the sea created distinct land areas that could be independently treated without fear of reinfestation. These 'rodent-proof glaciers' allowed the team to treat separate regions in different years. In the end, three phases of the project treated just under one-third of the island's 3500 square kilometres. The seals and many of the nesting birds either ignored the helicopters or took flight when they appeared.

Rodent activity stopped within a week of helicopter baiting. Researchers would find caches of bait the rats and mice had collected for storage. For the following six years, the researchers monitored for rat activity. Largely as expected, unfortunately, non-target mortality was observed in the form of bird kills. Carcasses of seven bird species, including skua, gulls, petrels, sheathbills and ducks, were discovered with evidence that they had died as a result of consuming bait pellets. However, populations of these birds appeared to recover over successive seasons, and researchers observed that 'although a substantial proportion of the local breeding

The rodent eradication from the 3500km² landscape of South Georgia was challenging, but was aided by the landscape. The glaciers and snow-covered mountains are inhospitable to rats and mice, and provided a barrier to their movement. *Tony Martin / South Georgia Heritage Trust*

skua population was probably lost to primary or secondary mortality, recruitment from outside the baited area quickly restored losses'. More pintail ducklings and sheathbills were seen than in many previous years.[3]

Because mice and several species of rat are super-tramps, going where people go, they have become a global environmental-extinction-causing scourge on islands. Thus, the eradication of mammals on islands is a tool increasingly used in conservation. A recent global review found 251 eradications on 181 islands around the globe.[4] The most common pests targeted were rodents (57%), goats (11%) and feral cats (8%). Across all the eradication programmes examined, 236 native species were documented to have subsequently benefited through increased population sizes or an increased distribution. Many more species almost certainly benefited but were not studied. Once free of the introduced pests, islands were recolonised or valued species were reintroduced. Birds were the most common beneficiaries, though others including reptiles and invertebrates benefited as well.

A South Georgia pintail (*Anas georgica georgica*) swimming in front of a male Southern elephant seal (*Mirounga leonina*) in South Georgia. The rodent eradication project resulted in a substantial by-kill of pintail, with populations projected to take <4 years to recover fully. The ducks are omnivorous and feed on seal carcasses. *Tony Heald / Nature Picture Library / Science Photo Library*

One example is an eradication of rats from Great Bird Island in the Caribbean Sea. This island is home to what is considered by many to be the only population of the world's rarest snake, the critically endangered Antiguan racer (*Alsophis antiguae*). After the rats were removed, the Antiguan racer population increased twentyfold. Other species on the island benefited too: the number of red-billed tropic birds doubled, and there was a dramatic increase in the breeding success of West Indian whistling ducks.[5]

But throughout the global review, just as on South Georgia, short-term negative impacts were observed. These included the death of gulls scavenging on rat carcasses in Alaska after a rat eradication programme. Although hundreds of native populations benefited from the pest management projects, it is of concern that eight populations of species were found to experience negative impacts after pest eradication. With more time, perhaps these populations will rebound. It is also possible that eradications can drive complex, indirect and unexpected ecological shifts in communities. These shifts include the eradication of one mammal releasing another from predatory control, resulting in this second species increasing its abundance effects on communities. While it is clear that these complex and unexpected effects can occur, the scorecard of eight negative

An Antiguan racer eating an anole lizard. The racer is one of the rarest and most endangered snakes in the world. It was extirpated from much of its former range by predation from the introduced mongoose and black rats. *John Cancalosi / Alamy*

effects versus 236 positive outcomes suggests we should continue pursing eradication as a conservation tool.

## Failed eradications: The ones that got away

Failed eradications can teach us a lot. And given that a large number of eradication attempts have failed all around the world, we have a lot to learn.

One of the very first recorded eradication attempts for an arthropod species was in 1890. It failed miserably. The target in question was the gypsy moth (*Lymantria dispar*), which was introduced into North America, specifically Massachusetts, by the French amateur entomologist Etienne Léopold Trouvelot. His exact motivations for the introduction are unclear, but he was probably trying to develop these moths into a silkworm industry.[6] Some of his gypsy moths escaped and reproduced, resulting in vast armies of very hungry caterpillars. Foliage feeding by these caterpillars led to the destruction of millions of trees throughout North America, with one 2011 study estimating them to cost US$868 million annually in damage.[7]

The methods used in initial attempts to eradicate the gypsy moth were crude: flame throwers burnt trees and caterpillars alike. Entire landscapes were coated with arsenic- and metal-based insecticides such as copper acetoarsenite and lead arsenate, the residues of which persist and are present in the soil today. This first eradication effort spanned 1890–1900 and had an estimated cost of US$1.2 million (equivalent to approximately US$28 million in 2010). It targeted a population that covered 42,500 square kilometres, encompassing a total of 30 towns and cities.[6] And it failed.

Caterpillar of a gypsy moth (*Lymantria dispar*).
*Didier Descouens / Wikimedia Commons*

Since then, many biodiversity managers have attempted gypsy moth eradication in many countries. With effective chemicals and application methods, gypsy moth incursions and infestations are frequently eradicated using time-tested approaches. On the west coast of the United States, gypsy moth incursions are monitored via pheromone traps. When individuals are observed, a standardised and proven eradication approach uses aerial and/or ground sprays of the biopesticide Btk (a specific control targeting lepidoptera, developed from the bacterial species *Bacillus thuringiensis* serotype *kurstaki*). Hundreds of hectares are sprayed, and gypsy moth populations are now frequently eliminated. The approaches, technology and tools have substantially improved to enable a high degree of confidence in dealing with gypsy moth incursions, with no records of failed attempts since 1984 (Fig. 4.2 and Fig. 4.3).

Many failed eradication attempts involve masses of money and wasted resources, and sometimes have nasty lingering after-effects. One example is the attempted eradication of the red imported fire ant (*Solenopsis invicta*) in

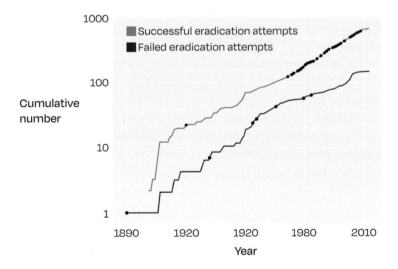

Fig. 4.2: The cumulative number of successful and failed arthropod eradication attempts over time. The dots are eradication attempts for the gypsy moth. Experience and technological developments have allowed biodiversity managers to become very capable in the eradication of this pest, with no failed eradication attempts since 1984. *GERDA*[11]

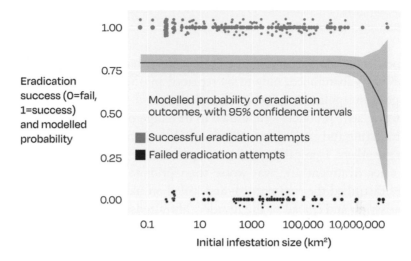

Fig 4.3. A statistical model of eradication success, which is dependent on the area infested. This analysis suggests that the probability of success is 72–84% for pests infesting <10 million km². The outcome of eradication attempts for pest infestations >10 million km² is less certain and highly variable, but still possible – though these attempts are likely to be expensive and resource intensive. *GERDA*[11]

the southern United States. This programme was cited as the 'largest and most devastating eradication program ever undertaken against an introduced insect pest'.[8]

Fire ants were first recognised as a new arrival to the United States in 1930. Native to South America, they spread from the port of Mobile in Alabama and occupied more than 93 million hectares in nine US states by 1989. Since then they have spread to at least 20 states. The first eradication attempt was in the state of Mississippi in 1948. A more widespread and intensive campaign began in 1957, with 10 cooperating states using the pesticides heptachlor and dieldrin at a rate of 2 pounds per acre over 20 million acres. Toxicologists Robert Metcalf et al. (1982) later commented on these chemicals:

> The insecticides chosen for control and later eradication efforts . . . can be collectively described as extremely persistent in the environment, broad-spectrum in their toxic action, highly bioaccumulative with long residence times in human and animal

tissues, and generally hazardous to a wide variety of terrestrial and aquatic wildlife. These insecticides have also been shown . . . to be chemical carcinogens.[9]

The pesticides killed mammals, birds, reptiles, fish and frogs. Eventually these pesticides were abandoned, only to be replaced by the chemical Mirex in 1962, with large-scale applications in Florida, Georgia and Mississippi in 1967. Mirex and its degradative products were shown to be dangerous to wildlife and humans, bioaccumulating and causing reproductive problems and nervous system disorders in factory workers. Ironically, the fire-ant problem in many areas bounced back quickly after treatment and was worse than prior to eradication attempts: the pesticides disturbed the environment and killed off native ant species, facilitating the colonisation and spread of the invader. Hundreds of millions of dollars have been spent on fire-ant eradication programmes over many decades without success; the ant is now more widespread and abundant in the United States than ever. Some scientists still debate the need for the eradication of the ant, as they consider it a relatively benign species causing just a little 'southern discomfort'.[8] Other scientists point to the ant being primarily a pest of human-disturbed or -altered environments that have low biodiversity.[10]

A shovel planted in a fire ant mound in an agricultural field in Texas. Worker ants are aggressively swarming over it in defence of their colony. Should any livestock stand on a mound, they will be attacked. There are around 30 mounds in this picture. *Phil Lester photo*

Even if local eradication is successful, biodiversity managers then have the challenge of keeping new incursions of pests out. For example, the successful rat eradication on snowy South Georgia that was discussed above could have been all for naught. Three years after eradication, rat footprints were seen in the snow after two ships berthed at a jetty.[3] The conservation group on the island swung into action to eliminate the new arrivals before they could become established.

Similarly, in New Zealand one famous Norway rat swam 400 metres of open ocean to an island from which its relatives had been previously eradicated. The marathon-swimming behaviour and subsequent movement of this rat on its new island could be closely followed because it wore a radio collar and had been genetically sequenced as part of another experiment. Still, it took more than four months of intensive work to finally trap and kill the elusive swimmer, perhaps highlighting the difficulty of eradication at extremely low population densities and when an individual displays atypical behaviours of bait- or trap-shyness.[11]

## What are the characteristics of successful eradications?

Wouldn't it be great to be able to predict whether an eradication attempt will be successful? A lot of money and effort could be saved and resources used more wisely if we could accurately predict management outcomes. In an effort to develop such a predictive ability, researchers have compiled large databases of eradication attempts, which can guide biodiversity managers as to what features, both of the organism itself and of the environment into which it has been introduced, are likely to predict success in eradication.

One such database is the Global Eradication and Response Database (GERDA), which has been developed to compile records of arthropod eradication programmes.[12] The authors of one 2014 publication drew upon 672 arthropod programmes from around the world. In their analysis, common species with multiple eradication attempts included the European gypsy moth (*Lymantria dispar*, with 73 cases), the Mediterranean fruit fly (*Ceratitis capitata*, with 56 cases), the oriental fruit fly (*Bactrocera dorsalis*, with 40 cases) and the yellow fever mosquito (*Aedes aegypti*, with 33 cases). Of those eradication attempts, 59% were successful, 16% failed and the remainder were either in progress or the results unknown. The key factors for predicting eradication success were found to include infestation size, method of detection, relative detectability of the organism, host range and primary feeding guild. The chances of successful eradication decreased with an increasing size of infestation. If pest organisms were easily detected, the chances of success were also higher. That is a logical result: if you can't find it, you can't kill it. And, interestingly, organisms that consumed a wide variety of food (polyphagous) were six times more likely to be eradicated than those with specialised diets. Note, however, that this high rate of eradication success of the species with polyphagous diets may not be directly related to the insects themselves. It may be that arthropods that attack a wide variety of plants are more aggressively targeted by government agencies due to their wide-ranging economic effects. The feeding mode or guild analysis indicated that insects such as those that feed within wood are much harder to eradicate, which is logical, as these are harder to find and thus kill.

What about the characteristics of successful eradication programmes for other pests? One European study examined factors that contribute to the success of campaigns to eradicate invasive alien invertebrates, plants and plant pathogens. Their dataset included 173 different campaigns, of which 51% were successful. Pathogens, bacteria and viruses were most likely to be eradicated, with fungi the least likely. For plants and invertebrates, the eradication probability was intermediate. The authors write:

> Our analysis indicates that initiating the campaign before the extent of infestation reaches the critical threshold, starting to eradicate within the first four years since the problem has been noticed, paying special attention to species introduced by the cultivation pathway, and applying sanitary measures can substantially increase the probability of eradication success.[13]

The sanitary measures they cite include management approaches such as banning the transfer of potentially contaminated material out of the infestation area. Like the GERDA database and analysis, the size of the infestation made a big difference, with the probability of successful eradication twice as high (67%) for infestations covering less than 5000 hectares than for eradication campaigns targeting larger areas (33%).

The size of the infestation, and people's attitudes, have influenced plant eradication programmes too. One review summarised the results of 30 plant eradication projects on the Galápagos Islands. Of the 30, only four were successful. While it is encouraging to know plant eradication is possible, all four successes covered less than one hectare of land with a single owner. The plants were species that did not have persistent seed banks. The authors of the review concluded that the 26 other projects failed because of a lack of support from institutions. There was no continuity of funding, and landowners also proved to be a problem when they denied permission to carry out the work.[14]

In New Zealand, 11 non-native plants have been eradicated. The common features of these were that all were found in just one or a few locations, and all were contained within discrete habitats. Philip Hulme describes three lessons that can be learned from New Zealand's attempts at plant eradications.[15] First, eradication is possible, but it requires evidence-based programmes that are nationally coordinated and backed by legislation. These programmes need financial support over several decades from a partnership with all stakeholders. Second, the stakeholders need to be engaged and contribute to the programme with surveillance, herbicide use and plant removal. Third and finally, while programmes in New Zealand are typically informed by a quantitative or semi-quantitative prioritisation analysis, these have invariably been over-optimistic.

Analysis of mammalian eradication attempts produced similar conclusions. The researchers who reviewed the work found that successful eradication

One of the larger eradication programmes was aimed at the coypu (*Myocastor coypus*). Native to South America, this semi-aquatic rodent can grow to weigh several kilograms. It was introduced to England in the 1920s for the fur trade and was eradicated from a 19,210 km² area over 1981–89, with more than 34,800 culled. *Basile Morin / Wikimedia Commons*

programmes require, but are not guaranteed by: (1) good planning, (2) a good pre-control study of the situation, (3) care in the selection of the eradication method or suite of methods, (4) sufficient financial and political support to complete eradication over the entire period of the programme and (5) public support.[16] An increasingly recognised feature of several successful mammalian eradication and restoration campaigns has been the involvement of the community in management decisions.[17, 18] Public engagement in all stages of eradication, from project conception to final evaluation, can yield public ownership and public pride. This ownership can foster ongoing biosecurity efforts to stop future reinvasion. Restoration is as much about the people as the natural environment, and the rights and input of Indigenous peoples are especially important. In some countries, Indigenous values have been enshrined in legislation. In New Zealand, recognition of the authority of Māori over land, water and biodiversity is critical in restoration and eradication programmes.[19]

It's worth noting that these characteristics of successful eradication programmes are not prescriptive. Eradication is still possible for pests that are hard to detect, that have been established for long periods and that have very large spatial distributions. These eradication programmes will typically require substantially more time and resources. Success is still possible, but costs for large eradication campaigns spanning many years can be immense (Fig. 4.4). And, occasionally, when a recent incursion is detected for a species that is recognised as problematic and that offers only a short window for management, governments will decide to act quickly, bypassing public consultation for large-scale management. Such a case occurred with the eradication of the black-striped mussel (*Mytilopsis sallei*) in Darwin, Australia.[20] This mussel is related to the invasive zebra mussel and is thought to be ecologically similar in its effects. The mussel was first observed and then confirmed on 27 and 29 March 1999, respectively. Between 3 and 22 April of the same year, 163,040 kilograms of liquid sodium hypochlorite and 4325 kilograms of powdered copper sulphate were added to one marina. Two other harbours were treated similarly. The addition of these chemicals will have caused a massive by-kill of many species, but the quick action was effective in eradicating this mussel. Had the government waited for a long period of consultation, the opportunity for eradication might have sailed away.

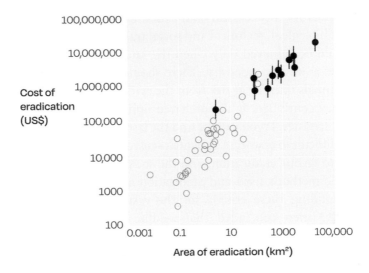

Fig. 4.4: The cost of mammalian pest eradications is proportionate to the area they occupy. The green open circles show eradications for mammal species on islands worldwide. The black closed circles show 11 large-scale invasive mammal eradications in Belgium, Great Britain and Ireland, with vertical bars indicating the range of cost estimates. These are based on equivalent USD in 2016. *Data and graph from Robertson et al. (2017)[46]*

## Attempt eradication, or use ongoing pest control?

Biodiversity managers typically view eradication as the most appealing management option for an undesired or invasive species. A perpetual freedom from the biodiversity, health or economic effects of the species is frequently seen as a much-preferred option in comparison to shouldering the burden of ongoing costs and management. But with the current available tools, technology and social support, eradication isn't always possible, and management might be the best option.

Imagine that a new species has arrived and become established: should we attempt an eradication, or just learn to live with and manage it via ongoing control programmes?

Mary Bomford and Peter O'Brien published an article in 1995 that asked just that question. When is the eradication of a pest feasible? They focused on vertebrate pest species, and suggested criteria that must be met and others that should be met for a successful eradication campaign. These criteria were a combination of biological, technical and social traits. The authors concluded that eradication could be achieved only when the rate of removal exceeded the rate of pest increase, at any pest density. There should also be no immigration. All reproductive animals must be at risk from the eradication methods implemented. Eradication, they continued, is the preferred option only when animals can be detected at low densities. If you cannot find the last few breeding pairs, an invasive species will be difficult to eradicate. A benefit–cost analysis should be implemented and be shown to favour eradication, and the socio-political environment should also support the methods used and eradication as an outcome. Because of the difficulty in fulfilling these criteria for the vast majority of invasive species, Bomford and O'Brien concluded that wildlife damage management should typically be based on ongoing control rather than eradication attempts that may be inappropriate for many invasive species.[21]

These criteria were developed for invasive vertebrate pests. More recently, a group of biodiversity managers suggested modifications and three additions to these criteria to incorporate invertebrate management approaches.[22] All nine are shown in Table 4.1. The criteria that were added were to note that small populations with limited spatial distributions are often easier to eradicate than those scattered over wider scales and have often been more successful. The authors also inserted a criterion to ensure that the environmental impacts of any eradication programme will be acceptable. These could include the presence of pesticide residues or non-target effects of their application, or the potential release of a second pest species from suppression by competition, predation or parasitism. The final additional criterion was in regard to the management of the programme. The authors noted that effective and meticulous planning and management, and clear lines of authority, are critical to successful pest eradication.[22] I think most people would argue that all nine criteria are appropriate for the incursion of virtually any microbe, plant or animal pests. Benefit–cost analyses, the ability to actually effect pest control, and public support are all important.

Typically, the information that we need in order to make an educated decision on attempts to eradicate a newly introduced species is incomplete. There is always uncertainty about some aspect of the programme, such as during the programme to eradicate the large white butterfly in New Zealand, whereby there was concern that immigration and emigration could not be prevented (see 'Case study 1: Eradication in action with large white butterflies', page 150).

| Criteria | Description |
|----------|-------------|
| 1. The pest population is forced to decline from one generation to the next, irrespective of density. | The removal rate needs to be more than the rate of replacement by reproduction, at any population density. An increase in resources, caused by reduced competition at low pest densities, may result in an increase in reproductive rates. Low densities are often harder and more expensive to find and remove. |
| 2. Every individual is able to be controlled at some stage of its development. | It is essential to stop individuals from reproducing, or to substantially reduce their reproductive rates. This could be achieved by culling reproductive adults, using fertility control, or stopping juveniles from becoming adults. |
| 3. Individuals can be detected at low population densities. | The decline of the pest population must be measurable. It is critical to be able to detect and control extremely low numbers of pests to ensure treatment efficacy and eventual eradication. |
| 4. Success is favoured by the small spatial extent of the population. | Several studies have found that when infested areas are small, eradication attempts are generally less expensive and more likely to succeed. Note that with changing technologies, larger pest populations have been eradicated. |
| 5. Immigration and emigration can be prevented. | If the target pest is able to immigrate into the eradication area or is released from captive populations, eradication is transient or unachievable. Immigration can be 'aided' by groups such as hunters re-introducing eradicated populations from controlled areas. |
| 6. Environmental impacts are acceptable. | Many control methods can have non-target effects. Biodiversity managers must consider if such impacts are socially and environmentally acceptable. Are the impacts reversible? Are they likely to be less than if the pest became established and widely distributed? |
| 7. Benefit–cost analysis favours eradication over control. | Benefit–cost analyses provide information that supports or opposes eradication attempts. These analyses can highlight how an initially large outlay could be cost-effective long term, and the economic benefit of the introduced species, such as attracting tourists for hunting. |
| 8. Suitable social, political, legal and institutional environment. | Are the control methods socially acceptable? Do managers have legal authority to implement control actions on public and private land? Will property owners allow pest eradication activities on their land? |
| 9. The programme is managed effectively. | Critical components for eradication success are to have efficient, meticulous and effective planning and management. Clear lines of authority are required, as well as access to expertise. |

Table 4.1: Suggested criteria for deciding if eradication is technically possible and preferable to continuing control for managing pests. All are deemed 'essential' except (3) and (7), deemed 'desirable'. *Adapted from Bomford & O'Brien (1995)[21] and Phillips et al. (2019).[22]*

It is unlikely that every single one of the criteria will be met in every pest-incursion scenario. For example, a detailed and accurate benefit–cost analysis is unlikely to be available for each pest in its new environment.

All the criteria in Table 4.1 are important. However, the big questions and hotly debated issues for eradication programmes are typically the methods and goals. Are the environmental impacts of the programme acceptable? Are such impacts likely to be less than those that would be sustained if the pest species became permanently established and more widely distributed? And will the control methods used be viewed as socially acceptable? We will deal with these questions in detail in a later chapter, though they are worth noting here. A programme involving the eradication of introduced possums in New Zealand likely needs to kill possums. There may be pain and suffering for the possums, which might be unacceptable to many people. The counter argument or perspective might be that the possums kill and cause pain and suffering for birds and other wildlife.

Some pest taxa represent distinct eradication challenges. For example, in New Zealand, bovine tuberculosis (bTB) is under an ongoing eradication campaign that has spanned three decades. One of the critical issues is that bTB can have many different hosts. The host species of primary concern is cattle, but other species, including possums, ferrets and hedgehogs, can carry infections. These are all introduced species and are generally recognised as pests themselves, with

A common brushtail possum and a ship rat attacking and eating birds from a nest in New Zealand. These mammals are clearly harmful to biodiversity. Their eradication would come with considerable benefits, but is it possible? *Ngā Manu Nature Reserve*

public support for their control as well. Consequently, the control of possums or ferrets in New Zealand is much less controversial and has higher levels of public support than the control of bTB in England. (See 'Case study 2: The eradication programme for bovine tuberculosis in New Zealand', page 152)

In Europe and England, native badgers host bTB. Badgers are renowned for their grumpy and aggressive dispositions, but nevertheless are afforded conservation under the Protection of Badgers Act 1992.[23] But this act and the passionate people seeking to conserve badgers create a problem for bTB management. To eliminate bTB in England, you'd need to eliminate badgers as a reservoir of bTB, either by killing them (which upsets many people) or by eliminating them as a host, perhaps by vaccinating badgers against the disease (which is hard). Obviously, the multi-host nature of this and many other pathogens can be a problem when making decisions around eradication.

In terms of the decision criteria represented in Table 4.1 for biodiversity managers, alternative hosts are represented by the criteria asking if the pest population can be forced to decline from one generation to the next, irrespective of its density. A similar approach for making stop/go decisions on eradication programmes was developed for mammals (Fig. 4.5). A key addition to this is the explicit questioning around the effects of the eradication on the biological community.

Left: Brian May (centre) is one of several celebrities who have added their voice to the campaign to stop the badger cull. An alternative to culling badgers is to vaccinate them. Right: A European badger is vaccinated against bovine tuberculosis during vaccination trials in Gloucestershire, 2011. *Left: Paul Davey / Alamy. Right: Neil Aldridge / Alamy*

*Does the introduced or invasive pest species:*
- fill an important ecological function no longer occupied by a native species?
- constitute the main prey for another introduced species, such that if eradicated, the predator will damage populations of indigenous species?
- limit the population of another introduced species, whose unchecked populations could lead to undesirable effects?
- allow, by its impact, the maintenance of desired communities?

 **WHETHER TO ACT**
NO　　　　　　　　　　　　　　　　　　　　　　YES

Are the likely results vague or uncertain?

Are the risks too high?

Is the probability of success too low?

 YES

No eradication. Reconsider management options.

 **HOW TO ACT**
NO

## ERADICATION

### Introduction history

Introduction date/s?

Origin of founding population/s?

Are there factors facilitating or enhancing the likelihood of invasion?

### Invaded community characteristics

By what mechanisms does the invasive species interact with indigenous species?

Does the invader have major consequences on the recipient community?

What is the nature and importance of the interaction between one or more species?

### Operation methods and conditions

What is the geographic scale of the eradication?

What is the best time of year to act for native and invasive species?

What is the ideal order and timing of operations?

Who should be consulted?

What is the best management option?

### Monitoring

What species should be monitored as a result of treatment?

How often should monitoring be done?

Can re-invasion be prevented?

Is scientific expertise available?

Is financial support available?

Is the probability of success too low?

Fig. 4.5: A decision diagram for eradication programmes. *Developed from Courchamp et al.*[16]

A wonderful example of how eradication programmes can have significant impacts via other community members comes from the northern California Channel Islands, where feral pigs were the target. Biodiversity managers on these islands had the goal of conserving and increasing populations of the critically endangered island fox *Urocyon littoralis*. However, golden eagles were attracted to the island by the abundant food source that the pigs provided and, once established, killed and ate foxes as well as pigs. Consequently, biodiversity managers knew they needed to remove or cull the eagles before removing the pigs. They contemplated shooting or lethal removal but, due to public controversy, did not implement this. Instead, prior to the successful lethal eradication of the pigs, they live-trapped or captured 44 eagle pairs with a net gun and helicopter. Unfortunately, one pair evaded capture and nested on the island. A diet analysis showed that this eagle pair intensified their consumption of foxes after pig eradication: fox represented 18% of the eagles' diets before pig removal, but 52% after pig culling. Clearly, if the threat of eagle predation had not been mitigated prior to pig removal, even more foxes would have been killed, possibly resulting in extinction. The study concluded that if complete eradication of all interacting species is not possible, the order in which pests or predators are removed requires careful consideration.[24]

A nestling golden eagle and her food on Santa Cruz Island. The nestling is believed to be from the pair that evaded capture. This nest contained the remains of 13 endangered island foxes, three of which had radio-telemetry collars.
*P. Sharpe, from Collins et al. (2009)[24]*

## The increasing scale of successful eradications: How big can we go?

There is no doubt that we are getting better and better at pest eradication. We've seen new technologies and approaches used on extremely large islands such as South Georgia. The spatial scale of successful rodent eradications has increased by an order of magnitude each decade from 1964 to 2004 in New Zealand.[25]

One of the largest, most audacious and most ambitious eradication plans of recent decades is the New Zealand government's Predator Free 2050 programme. The conception of this programme has largely been credited to the physicist and science communicator Sir Paul Callaghan, who in 2012 suggested New Zealand could save the nation's fauna with the 'crazy and ambitious' plan to eradicate invasive pests.[26] The goal of the programme is to eradicate seven of the most damaging introduced predators from all of New Zealand by 2050. The targets are three species of rats (*Rattus* spp.), three species of mustelids (*Mustela* spp.) and brushtail possums (*Trichosurus vulpecula*). These species were chosen, collectively, because 'they inflict the worst damage of all the introduced predators on New Zealand's wildlife. We also know more about their biology and control than any other predators.'[27]

Largely as a consequence of the ambitious scale of the proposed eradication, the Predator Free 2050 programme has many critics. Some have suggested that 'the policy appears to be flawed on multiple levels: technical, financial, social, ecological, and ethical' and is 'diverting effort and resources from higher environmental priorities and better alternatives'. These authors would prefer conservation efforts that are less 'bombastic'.[28] Other commentators object to government agencies and the supporting predator-free organisations in New Zealand using the rhetoric of 'war', 'invasion' and 'military' strategy.[29] Indeed, the war-like terminology and ethos have resulted in some extreme behaviours, such as in 2017, at a school fundraiser, where students drowned juvenile possums (joeys) and humiliated possum corpses. Moreover, the complete eradication of pests using current techniques would be extremely costly and requires massive investment. Dr Bruce Warburton, of the government-owned research institute Manaaki Whenua – Landcare Research, said at a recent conference that eradicating these seven pests by 2050 was 'just a dream'. He continued, 'There's a huge difference between a 98% kill and a 100% kill. A 95% to 98% kill with 1080 operations costs NZ\$30–\$40 a hectare, but a 100% kill and total eradication would cost NZ\$200,000 to \$500,000 a hectare.'[30] This difference between pest control and total eradication is huge, at least with current techniques and approaches for pest management.

It is clear that new eradication and pest management tools will be needed to achieve New Zealand's goal of becoming predator free by 2050. A 2020 analysis examined the likelihood of rat eradication success on some New Zealand islands and how long it might take to achieve this result for others using currently available techniques. The authors included a range of factors that influence eradication including the island size, the proximity of the island to the mainland, whether it is under public or private ownership, whether it is populated by humans and whether rat eradication had already been achieved. Overall, with our current technologies and pest management approaches, only 14 out of 74 islands were modelled as likely to be rat-free by 2050. The authors conclude that 'it is imperative for New Zealand to invest in, and develop, novel technical and social tools for eradication to increase the current rate of eradication successes'.[31] The buy-in or social acceptability of these tools will be especially important in islands or regions with high human populations. Clearly, new approaches for pest control will be needed if we are to have any hope of eradicating these seven pest species from New Zealand.

## Costs, communication and the Lazarus effect

Eradications cost a considerable amount of money. Earlier in this chapter we looked at the ongoing programme for eradication of the cattle disease *Mycoplasma bovis* in New Zealand. That programme has been estimated to cost nearly NZ$1 billion. Estimates for the eradication of the seven species in the Predator Free 2050 programme are as high as NZ$32 billion, equating to 0.5% of the nation's annual gross domestic product (GDP).[28] Even the relatively small eradication of the large white butterfly was estimated to cost nearly NZ$5 million. Unfortunately, the substantial costs and technical expertise involved with elimination or eradication mean that they are prohibitive for many developing nations around the globe, which is especially problematic when you consider how many developing nations are global biodiversity hotspots where we most need conservation. Further, the inability of developing nations to eradicate or control invasive pests can then become their neighbours' problems, as pest numbers grow and disperse.

The large white butterfly (*Pieris brassicae*) posed a major threat to native plant species in New Zealand. *Damian Waters / Alamy*

Cost and public perception of expenditure often become major issues near the end of eradication programmes, when pest numbers have declined and the problems they cause are typically out of the public eye. There is often publicity exhaustion wherein people are sick of hearing about this pest ('Haven't we done that already? Why are we still spending money on that eradication when there are other more pressing problems?'). This sort of public exhaustion has been termed 'campaign fatigue' and was a major issue with the successful global polio eradication programme.

Key lessons from the polio programme are relevant to eradication campaigns everywhere. Social mobilisation and community participation were vital in polio eradication, especially from within communities, rather than from a government. As Obregon and Waisbord (2010) write: 'Families were persuaded and convinced by the teams of interns, social workers and influential persons that polio drops did not have any side effects. They were more receptive to the advice given by medical interns compared to other staff members of the Government District Hospital because of quality of health services provided to the community.'[32] Thus, a cookie-cutter approach of applying the same communication strategy everywhere, for any eradication programme, is unlikely to succeed. An appropriate communication plan and social mobilisation should be informed by local distinctiveness, perceptions, attitudes and structures; the dynamics of local power and participation; and the role of local influencers. An effective, ongoing,

A painting of the musk shrew (*Suncus murinus*) by British artist
Olivia Fanny Tonge. *Natural History Museum, London / Science Photo Library*

locally oriented communication plan is probably the key to maintaining public
support and engagement for any eradication programme.

Finally, when should you call an eradication programme successful? Ending
a programme too soon can mean that not all individuals are killed or removed,
leading to the Lazarus effect: the sudden reappearance of a species thought to
be extinct or eradicated.[33] A well-examined example is the failed attempt to
eradicate the invasive Asian musk shrew (*Suncus murinus*) from a 25-hectare
island nature reserve in the Indian Ocean. Live traps were used in the eradication
programme, which was pronounced a success after four intensive trapping
sweeps encompassing 76 days. Six months after the eradication campaign ended,
however, shrews began to reappear and became abundant. Eradicated, they
were not. Authors analysing this programme assessed the potential role of low
efficiency of all traps, or low trap efficiency in specific locations that provided a
refuge in some areas of the island, or the trap-shyness of some individuals. They
concluded that trap-shy behaviour was likely to be responsible.[34] This is a common
outcome: other researchers have similarly concluded that no single technique can
be relied on to detect all individuals, and often no single eradication technique
will kill all individuals within a population.[35]

The final decision as to when to declare success and stop an eradication
programme is difficult and must incorporate specific aspects of the organism's
biology and ecology. It is impossible to generalise about a specific time period

for all organisms. For example, I think we could quickly assess thc success of a programme to eradicate elephants from a 25-hectare island, but our confidence in quickly finding every last small Asian musk shrew would be much lower. The eradication of even large mammals such as goats can be challenging and expensive, and it requires certainty before declaring success (see 'Case study 3: Goats on Galápagos', page 153).

## Case study 1: Eradication in action with large white butterflies

The large white butterfly, *Pieris brassicae*, was first detected in New Zealand in May 2010 in Nelson.[36,37] This butterfly is native to the Palearctic region of the northern hemisphere but has established in South Africa and South America. It probably travelled to New Zealand as pupae in shipping containers. The butterfly has a wide host diet and is known from its native range to eat at least 91 different species from 12 different plant families. It specialises in the Brassicaceae family, which are mustards, the crucifers, and the cabbages. In New Zealand the native Brassicaceae comprise approximately 3% of indigenous flora. The conservation status of over half of these species is 'threatened' or 'nationally critical'. Due to

Entomologist Richard Toft searches for large white butterflies in Nelson, with caterpillars on a leaf in the foreground.
*Tim Cuff / Alamy*

its invasive nature and broad host range, the large white butterfly was listed as an 'unwanted organism' under the New Zealand Biosecurity Act 1993. MPI, the agency responsible for implementing this act, responded by surveying the region, monitoring and evaluating the potential for eradication. In 2012 MPI terminated its response, because it considered an eradication attempt would probably fail. They also predicted that the expected benefit-to-cost ratio for an eradication was too small.

The Department of Conservation Te Papa Atawhai had a different view. This government agency has a responsibility to protect native biodiversity under the Conservation Act 1987. In November 2012 they initiated an eradication attempt based on the risks the butterfly posed to native plants. This involved encouraging people to search for the butterfly and report sightings. A bounty was offered over a short period. Host plants were searched by trained employees, and ground-based spraying of a BioGro-certified organic insecticide was carried out to kill eggs and larvae. Populations of natural enemies, such as parasitic wasps known to attack large white butterflies, were augmented in the region. The last detection of the large white butterfly was 16 December 2014 – and the butterfly was officially declared eradicated from New Zealand on 22 November 2016. It was the first time a butterfly had been eradicated.[37] The total cost was estimated at NZ$4.107 million.[36]

Why did DOC and MPI respond differently to the same incursion? The pest management priorities in each organisation were different, and money in both organisations is limited. MPI evaluates eradication by calculating a benefit cost ratio (BCR) using the pest's likely impact over a 20-year period, the predicted cost of an eradication attempt, and the eradication success probability. For MPI to initiate an eradication attempt, a BCR over 3:1 is required, which was not met in this case. There are, however, many unknowns in this and most similar analyses. What is the likelihood of success? What value do we place on the conservation of native species? DOC used the modified criteria shown in Table 4.1 for deciding whether eradication was technically possible and preferable to continuing control for managing pests, which were developed for this eradication programme. Nevertheless, doubt persisted that immigration and emigration could be prevented (see (5) on Table 4.1). Perhaps the hilly topography of the area and abundance of introduced predatory wasps (*Vespula* and *Polistes spp.*) helped with its success.

## Case study 2: The eradication programme for bovine tuberculosis in New Zealand

*Mycobacterium bovis* is a bacterial pathogen with a global distribution. It is the causative agent of bovine tuberculosis (bTB), which infects a broad range of hosts including humans, brushtail possums (*Trichosurus vulpecula*), deer (*Cervus spp.*), ferrets (*Mustela furo*), feral pigs (*Sus scrofa*), and hedgehogs (*Erinaceus europaeus*). The exact date of bTB introduction into New Zealand is unknown, but as early as 1880 the disease was well established: up to 7% of cattle slaughtered in Wellington were considered infected. Domestic pigs had even higher rates of infection, most likely due to being fed waste milk products. Until 1956, bTB testing in cattle was voluntary and the control of this disease was ineffective. The government introduced legislation so that by 1970 all cattle were under compulsory testing and infected animals had to be culled. These policies led to a substantial reduction in infection rates with the exception of some regions, such as the upper West Coast of the South Island, where, in 1971, 82% of herds were thought to be infected.[38]

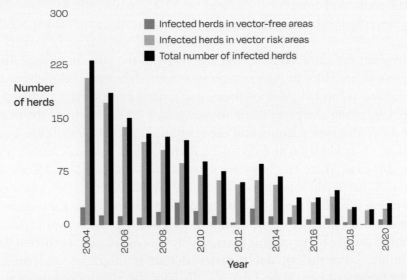

Fig 4.6: The number of cattle herds in New Zealand infected by bTB has been declining, indicating that the controls and eradication programme may be working. There are, however, still infections occurring, even in 'Vector Free Areas' where possum control is undertaken.

To understand the high infection rates on the West Coast, a small experiment was performed. A herd of just 29 calves was grazed in a paddock known to be in an area with possums that had high rates of bTB infections. After six months, 16 of the 29 calves were infected and had 'gross lesions'. Possums were implicated as disease spreaders. That small experiment has gone on to influence bTB management around the world.

A following experiment culled possums, resulting in a drop of bTB infections in the local cattle population from 16% to 4% after three years.[39] The possums interact with cattle by foraging, urinating and defecating in the same fields, or by inquisitive cattle displaying an interest in dead possum carcasses.

bTB is now under an eradication plan in New Zealand, which is enforced under the Biosecurity Act 1993, called the National Bovine Tuberculosis Pest Management Plan.[40] The plan costs approximately NZ$80 million per year and is funded by both the government and groups including dairy farmers. Key management practices involve the testing and management of diseased cattle and deer, movement restrictions of infected herds and from areas with high bTB infections, and ground and aerial operations to reduce pest numbers (primarily possums) which carry and spread TB to farmed animals. Rates of infection have dropped dramatically under this eradication plan. So far, bTB appears to have been eradicated from nearly 1.6 million hectares, and the number of infected herds has reduced to 43. The goals are now to eradicate bTB from cattle and deer herds by 2026, from possums by 2040, and from New Zealand by 2055.[41]

This case highlights the role that alternative hosts, and their management, can play in disease dynamics and eradication. Statistical analysis and modelling have predicted that possum control will eradicate TB from possum–deer–pig host communities in forest habitats. But, in grassland ecosystems, TB is predicted to persist in the ferret–pig host complex, even in the absence of possums, potentially jeopardising the effectiveness of possum-only control programmes.[42]

## Case study 3: Goats on Galápagos

The Galápagos has been described as the 'cradle of our understanding of evolution'.[43] Unfortunately, since the discovery of this volcanic archipelago, it has been the recipient of over 1400 introduced species. New residents include 540 insect species and 36 introduced vertebrates. Many of the introduced species

A finch chick with *Philornis downsi* fly parasites. The chick will likely die or have deformities as an adult. *Jody O'Connor*

occur in low densities and are not considered pests or problems; others are of major concern. For example, the avian parasitic fly *Philornis downsi* has larvae that are obligatory parasites of bird chicks, including Darwin's finches. One estimate of the average mortality of fly-infected chicks was 76%; in some years it reaches 100%, with surviving nestlings often deformed and anaemic. Biodiversity managers would love to see the fly eradicated from the Galápagos but lack the necessary tools.

Eradication has been attempted for several other invasive species on the Galápagos Islands, and some attempts have failed due to inefficient resources or other difficulties. One of the first programmes, however, was simple and successful: the eradication of five goats from the island of South Plaza in 1961. A much larger number of goats was eradicated from the 58,465-hectare Santiago Island in 2001–07.[44] There, goat herbivory had halted growth and recruitment of trees, cacti and forbs. Overgrazing had led to soil erosion and was a major contributing factor to the extinction of at least one plant, the Galápagos amaranth (*Blutaparon rigidum*).

More than 79,000 goats were removed from Santiago Island, costing approximately US$4.5 million. A suite of ground and aerial hunting techniques were used. Initially, ground-hunting techniques with dogs removed many of the goats. Hunters then began using 'Judas' goats and 'Mata Hari' goats. The Judas goats were those that had been captured from nearby islands, radio-collared, then released. Goats are gregarious, so these released individuals would join others, aiding hunters to locate remnant populations. The Mata Hari goats were female, sterilised Judas goats, with hormone implants that made them more effective at attracting the male goats. (Mata Hari was the stage name of a Dutch dancer who was convicted of being a spy for Germany during WWI. Unbound and without a blindfold, she was said to have blown a kiss to the French firing squad who executed her in 1917.)

The last individuals in many populations are especially hard to find. Killing the last 1000 goats on Santiago Island was estimated to have cost US$2 million. Then, an additional US$467,064 was spent on monitoring to confirm the eradication.[44] This monitoring and confirmation was considered especially important given that goat eradication was falsely declared on other islands, only for populations to rebound and recover.[45] Another issue has been with fishermen releasing goats to islands where previously they had been eradicated.

Feral goats (*Capra sp.*) overlooking the Alcedo volcanic crater on Isabela Island of the Galápagos archipelago. These goats were introduced by human settlers and eat a wide diversity of plants. Giant tortoises and land iguanas have seen their populations decimated since the arrival of these adaptable creatures.

*Science Photo Library / Alamy*

## Further reading and discussion

1. In 2001 an unidentified marine fouling pest, smothering wharf piles and moorings in a northern harbour, was recorded for the first time in New Zealand. An eradication programme was investigated but not implemented. Was that the right decision?

*See:* Coutts, A. D. M. & Forrest, B. M. (2007). Development and application of tools for incursion response: Lessons learned from the management of the fouling pest *Didemnum vexillum*. *Journal of Experimental Marine Biology and Ecology, 342* (1), 154–162. doi.org/10.1016/j.jembe.2006.10.042

2. The goal of making New Zealand predator free by 2050 has drawn both criticism and praise. The authors of one review suggest that achieving this goal is not a simple 'scaling up' of current eradication efforts. They highlight key impediments and issues regarding eradicating invasive species at a national scale. Which do you think is the critical problem for the Predator Free 2050 programme?

*See:* Peltzer, D. A., et al. (2019). Scale and complexity implications of making New Zealand predator-free by 2050. *Journal of the Royal Society of New Zealand, 49*(3), 412–439. doi.org/10.1080/03036758.2019.1653940

Carpet sea squirt (*Didemnum vexillum*), also known as 'sea vomit'. It attaches to hard surfaces and can look like a yellowish wax dripping over a structure. *Woods Hole Coastal and Marine Science Center / USGS*

3. As human populations and pressure on the world grow, we see higher rates of pest movement and biological invasions. Rates of invasion and biodiversity extinctions will increase over the next thousand years and more. In an evolutionary timescale and framework, are we wasting our time and resources by attempting pest eradication from islands for conservation benefit?

*See:* Capinha, C., et al. (2015) The dispersal of alien species redefines biogeography in the Anthropocene. *Science, 348*(6240), 1248–1251. doi. org/10.1126/science.aaa8913

4. An argument has been made that many invasive species function as 'replacements' for biodiversity that has been driven extinct by humans. Introduced hippos replace extinct *Toxodon* in South America, or goats might replace giant tortoises as large herbivores on the Galápagos Archipelago. Are replacement species such as goats really the same as giant tortoises? What do studies tell us about biodiversity impacts after invasive species eradication programmes?

*See:* Bastille-Rousseau, G., et al. (2017). Ecosystem implications of conserving endemic versus eradicating introduced large herbivores in the Galápagos Archipelago. *Biological Conservation, 209*: 1–10. doi.org/10.1016/j. biocon.2017.02.015

A Galápagos giant tortoise. *Phil Lester photo*

# References

1. Bellingham, P. J., Towns, D. R., Cameron, E. K., et al. (2010). New Zealand island restoration: Seabirds, predators, and the importance of history. *New Zealand Journal of Ecology, 34*(1), 115–136. newzealandecology.org/nzje/2905

2. Corona, A. (2020, March 6). Disease eradication: What does it take to wipe out a disease? *American Society for Microbiology.* asm.org/Articles/2020/March/Disease-Eradication-What-Does-It-Take-to-Wipe-out

3. Martin, A. R., & Richardson, M. G. (2017). Rodent eradication scaled up: Clearing rats and mice from South Georgia. *Oryx, 53*(1), 27–35. doi.org/10.1017/s003060531700028x

4. Jones, H. P., Holmes, N. D., Butchart, S. H. M., et al. (2016). Invasive mammal eradication on islands results in substantial conservation gains. *Proceedings of the National Academy of Sciences, 113*(15), 4033–4038. doi.org/10.1073/pnas.1521179113

5. Daltry, J. C., Bloxam, Q., Cooper, G., et al. (2009). Five years of conserving the 'world's rarest snake', the Antiguan racer *Alsophis antiguae. Oryx, 35*(2), 119–127. doi.org/10.1046/j.1365-3008.2001.00169.x

6. Tobin, P. C., Bai, B. B., Eggen, D. A., et al. (2012). The ecology, geopolitics, and economics of managing *Lymantria dispar* in the United States. *International Journal of Pest Management, 58*(3), 195–210. doi.org/10.1080/09670874.2011.647836

7. Aukema, J. E., Leung, B., Kovacs, K., et al. (2011). Economic impacts of non-native forest insects in the continental United States. *PLoS One, 6*(9), Article e24587. doi.org/10.1371/journal.pone.0024587

8. Davidson, N. A., & Stone, N. D. (1989). Imported fire ants. In D. L. Dahlsten, R. Garcia, & H. Lorraine (eds.), *Eradication of exotic pests: Analysis with case histories* (196–217). Yale University Press. doi.org/10.2307/j.ctt2250vh8.16

9. Metcalf, R. L., Severn, D. L., Alley, E. L., et al. (1982.) Environmental toxicology. In S.L. Battenfield (ed.), *Proceedings of the symposium on the imported fire ant*, June 7–10, Atlanta, Georgia, 75-8r.

10. King, J. R., & Tschinkel, W. R. (2016). Experimental evidence that dispersal drives ant community assembly in human-altered ecosystems. *Ecology, 97*(1), 236–249. doi.org/10.1890/15-1105.1

11. Russell, J. C., Towns, D. R., Anderson, S. H., et al. (2005). Intercepting the first rat ashore. *Nature, 437*(1107). doi.org/10.1038/4371107a

12. Tobin, P. C., Kean, J. M., Suckling, D. M., et al. (2014). Determinants of successful arthropod eradication programs. *Biological Invasions, 16*(2), 401–414. doi.org/10.1007/s10530-013-0529-5

13. Pluess, T., Jarosik, V., Pyšek, P., et al. (2012). Which factors affect the success or failure of eradication campaigns against alien species? *PLoS One, 7*(10), Article e48157.

doi.org/10.1371/journal.pone.0048157

14. Gardener, M. R., Atkinson, R., & Rentería, J. L. (2010). Eradications and people: Lessons from the plant eradication program in Galapagos. *Restoration Ecology, 18*(1), 20–29. doi.org/10.1111/j.1526-100X.2009.00614.x

15. Hulme, P. E. (2020). Plant invasions in New Zealand: Global lessons in prevention, eradication and control. *Biological Invasions, 22*, 1539–1562. doi.org/10.1007/s10530-020-02224-6

16. Courchamp, F., Chapuis, J. L., & Pascal, M. (2003). Mammal invaders on islands: Impact, control and control impact. *Biological Reviews, 78*(3), 347–383. doi.org/10.1017/s1464793102006061

17. Oppel, S., Beaven, B. M., Bolton, M., et al. (2011). Eradication of invasive mammals on islands inhabited by humans and domestic animals. *Conservation Biology, 25*(2), 232–240. doi.org/10.1111/j.1523-1739.2010.01601.x

18. Peltzer, D. A., Bellingham, P. J., Dickie, I. A., et al. (2019). Scale and complexity implications of making New Zealand predator-free by 2050. *Journal of the Royal Society of New Zealand, 49*(3), 412–439. doi.org/10.1080/03036758.2019.1653940

19. Norton, D. A., Young, L. M., Byrom, A. E., et al. (2016). How do we restore New Zealand's biological heritage by 2050? *Ecological Management & Restoration, 17*(3), 170–179. doi.org/10.1111/emr.12230

20. Willan, R. C., Russell, B. C., Murfet, N. B., et al. (2000). Outbreak of *Mytilopsis sallei* (Récluz, 1849) (Bivalvia: Dreissenidae) in Australia. *Molluscan Research, 20*(2), 25–30. doi.org/10.1080/13235818.2000.10673730

21. Bomford, M., & O'Brien, P. (1995). Eradication or control for vertebrate pests? *Wildlife Society Bulletin, 23*(2), 249–255. jstor.org/stable/3782799

22. Phillips, C. B., Brown, K., Broome, K., et al. (2019). Criteria to help evaluate and guide attempts to eradicate terrestrial arthropod pests. *Occasional Paper of the IUCN Species Survival Commission, 62*, 400–404. portals.iucn.org/library/node/48358

23. Protection of Badgers Act 1992. legislation.gov.uk/ukpga/1992/51/contents

24. Collins, P. W., Latta, B. C., & Roemer, G. W. (2009). Does the order of invasive species removal matter? The case of the eagle and the pig. *PLoS One, 4*(9), Article e7005. doi.org/10.1371/journal.pone.0007005

25. Clout, M. N., & Russell, J. C. (2006). The eradication of mammals from New Zealand islands. In F. Koike, M. N. Clout, M. Kawamichi, M. De Poorter, & K. Iwatsuki (eds.), *Assessment and Control of Biological Invasion Risks* (127–141). International Union for Conservation of Nature and Shoukadoh Book Sellers. stat.auckland.ac.nz/~jrussell/files/papers/CloutRussell2006.pdf

26. Owens, B. (2017). Behind New Zealand's wild plan to purge all pests. *Nature, 541*, 148–150. doi.org/10.1038/541148a

27. Department of Conservation. (2020). *Towards a predator free New Zealand: Predator Free 2050 Strategy* (978-0-473-51291-0). doc.govt.nz/globalassets/documents/conservation/threats-and-impacts/pf2050/pf2050-towards-predator-freedom-strategy.pdf

28. Linklater, W., & Steer, J. (2018). Predator Free 2050: A flawed conservation policy displaces higher priorities and better, evidence-based alternatives. *Conservation Letters, 11*(6). doi.org/10.1111/conl.12593

29. Morris, M. C. (2019). Predator free New Zealand and the 'war' on pests: Is it a just war? *Journal of Agricultural and Environmental Ethics, 33*(1), 93–110. doi.org/10.1007/s10806-019-09815-x

30. Woolf, A.-L. (2019, May 21). Predator Free 2050 'dream' called too costly, too unlikely at Biological Heritage conference. *Stuff.* stuff.co.nz/environment/112879315/predator-free-2050-dream-called-too-costly-too-unlikely-at-biological-heritage-conference

31. Carter, Z. T., Lumley, T., Bodey, T. W., et al. (2020). The clock is ticking: Temporally prioritizing eradications on islands. *Global Change Biology, 27*(7), 1443–1456. doi.org/10.1111/gcb.15502

32. Obregon, R., & Waisbord, S. (2010). The complexity of social mobilization in health communication: Top-down and bottom-up experiences in polio eradication. *Journal of Health Communication, 15, Suppl 1*, 25–47. doi.org/10.1080/10810731003695367

33. Flessa, K. W., & Jablonski, D. (1983). Extinction is here to stay. *Paleobiology, 9*(4), 315–321. doi.org/10.1017/S0094837300007776

34. Seymour, A., Varnham, K., Roy, S., et al. (2005). Mechanisms underlying the failure of an attempt to eradicate the invasive Asian musk shrew *Suncus murinus* from an island nature reserve. *Biological Conservation, 125*(1), 23–35. doi.org/10.1016/j.biocon.2005.03.005

35. Morrison, S. A., Macdonald, N., Walker, K., et al. (2007). Facing the dilemma at eradication's end: Uncertainty of absence and the Lazarus effect. *Frontiers in Ecology and the Environment, 5*(5), 271–276. doi.org/10.1890/1540-9295(2007)5[271:Ftdaee]2.0.Co;2

36. Brown, K. et al. (2019). Feasibility of eradicating the large white butterfly (*Pieris brassicae*) from New Zealand: Data gathering to inform decisions about the feasibility of eradication. In R. Veitch, M. N. Clout, A. R. Martin, J. C. Russell & C. J. West (eds.), *Island invasives: Scaling up to meet the challenge, 62* (364–369). IUCN Species Survival Commision.

37. Phillips, C. B., Brown, K., Green, C., Toft, R., Walker, G., Broome, K. (2020). Eradicating the large white butterfly from New Zealand eliminates a threat to endemic Brassicaceae. *PLoS One 15*(8), e0236791. doi.org/10.1371/journal.pone.0236791

38 Livingstone, P. G., Hancox, N., Nugent, G. & de Lisle, G. W. (2015, March 25). Toward eradication: The effect of *Mycobacterium bovis* infection in wildlife on the evolution and future direction of bovine tuberculosis management in New Zealand. *New Zealand Veterinary Journal 63* (suppl 1), 4–18. doi.org/10.1080/00480169.2014.971082

39  Davidson, R. M. (1976). The role of the opossum in spreading tuberculosis. *New Zealand Journal of Agriculture, 133*, 21–25.

40. Biosecurity (National Bovine Tuberculosis Pest Management Plan) Order 1998. legislation.govt.nz/regulation/public/1998/0179/latest/whole.html.

41. OSPRI. (2022). About the TBfree programme. ospri.co.nz/our-programmes/the-tbfree-programme/about/

42. Barron, M. C., Tompkins, D. M., Ramsey, D. S. & Bosson, M. A. (2015). The role of multiple wildlife hosts in the persistence and spread of bovine tuberculosis in New Zealand. *New Zealand Veterinary Journal, 63*(suppl 1), 68–76. doi.org/10.1080/00480 169.2014.968229

43. Moore, R. & Cotner, S. (2014). *Understanding Galápagos: What you'll see and what it means*. McGraw Hill.

44. Cruz, F., Carrion, V., Campbell, K. J., Lavoie, C. & Donlan, C. J. (2009). Bio-economics of large-scale eradication of feral goats from Santiago Island, Galápagos. *Journal of Wildlife Management, 73*(2), 191–200. doi.org/10.2193/2007-551

45. Campbell, K., Donlan, C.J., Cruz, F. & Carrion, V. (2004). Eradication of feral goats *Capra hircus* from Pinta Island, Galápagos, Ecuador. *Oryx, 38*(3), 328–333. doi.org/10.1017/S0030605304000572

46. Robertson, P. A. et al. (2017). The large-scale removal of mammalian invasive alien species in Northern Europe. *Pest Management Science, 73*(2), 273–279. doi.org/10.1002/ps.4224

# 5. PEST CONTROL

## Chemicals and pesticides, biological control and other means of killing or controlling

Perhaps the most polarising and controversial method of killing pests in New Zealand is the use of a toxin called 1080. New Zealand has been estimated to use about 80% of the global supply of 1080, or sodium fluoroacetate, to control populations of possums, rabbits, rats and stoats. In 2011 New Zealand was estimated to be using 3000 kilograms of sodium fluoroacetate annually.[1] At that time, Australia was the next biggest employer of 1080, using 200 kilograms annually for the control of introduced foxes and rabbits. The primary reason for the relatively high use in New Zealand is that there are no native mammals here, other than bats, that concern conservation authorities as potential non-target by-kill. The Department of Conservation Te Papa Atawhai had the goal of applying 1080 to approximately 474,000 hectares of conservation land in the 2020/2021 application season, equating to nearly 6% of New Zealand's total conservation

Introduced in 1884 for rabbit control, the stoat (*Mustela erminea*) is a significant pest in New Zealand. *Steve Hillebrand, USFWS / Wikimedia Commons*

estate.[2] Additionally, 1080 is used by OSPRI Ltd in the campaign to rid New Zealand of bovine tuberculosis over approximately 400,000 hectares.[1]

In 2007 the New Zealand Environmental Risk Management Authority completed what they described as 'the largest and most challenging exercise ever undertaken' by their organisation – a reassessment of 1080 use.[3] They received more than 1400 submissions spanning the full spectrum of possible views. Conservation groups, regional councils and many farming groups argued 1080 was an essential tool – imperfect, but necessary, given the lack of any practicable alternatives. Without 1080, bird extinctions were highly likely in the near future, or bovine tuberculosis would become rampant and devastate the dairy and cattle industry. Opponents of 1080 use included many deerstalkers and pig-hunters. They spoke about the introduced deer and wild pig populations that are killed as a by-product of 1080 applications. Some had farm animals or dogs that had been killed by 1080. Animal welfare groups also voiced concern over the painful deaths that animals can suffer from this poison. Other submitters worried that

Stoat predation has been identified as one of the primary causes of the extinctions of the huia, laughing owl, New Zealand little bittern, South Island piopio and South Island kōkako. On the left, a North Island kōkako. On the right, a kea. *Left: Matt Binns / Wikimedia Commons. Right: Michal Klajban / Wikimedia Commons*

1080 was a threat to human health or that the toxin was killing the native species it was dropped to protect. Presentations from Māori representatives similarly spanned the spectrum, from strong opposition to the use of 1080 to support for its continued use as an effective pest management tool.

The conclusion of the 2007 reassessment of 1080 was 'that the positive effects (benefits) of the substances outweigh the adverse effects (risks and costs)'. This enabled the continued use of 1080, albeit with additional safety measures. It didn't, however, stop the debate. Another comprehensive review of this pesticide came in 2011 from the Parliamentary Commissioner for the Environment.[1] With all the protests regarding 'widespread' and 'indiscriminate' 1080 use, the commissioner expressed surprise that 1080 was used so little. Further extinctions were considered likely because only 6% of the conservation estate was being treated (Fig. 5.1). The commissioner concluded that, 'based on careful analysis of the evidence . . . not only should the use of 1080 continue (including in aerial

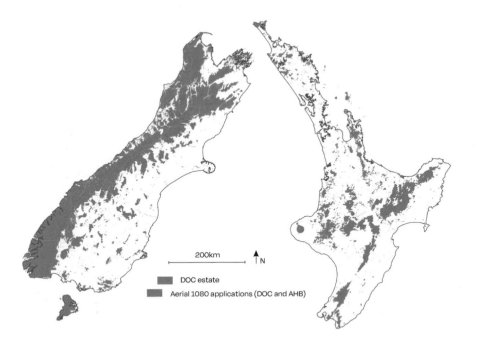

200km　　↑N

DOC estate

Aerial 1080 applications (DOC and AHB)

Fig. 5.1: The orange sections show the areas where DOC and the Animal Health Board carried out aerial drops of 1080 from July 2008 to June 2009. *Adapted from Parliamentary Commissioner for the Environment (2011)*[1]

operations) to protect our forests, but . . . we should use more of it'.[1] (See 'Case study 1: Evaluating the use of 1080', page 200) The evidence of the positive effects of 1080 applications for the conservation of New Zealand's biodiversity is consistent and overwhelming (Fig. 5.2). Nevertheless, anti-1080 groups and feelings are still widespread in New Zealand. DOC staff are regularly threatened or abused, and legal action attempts to stop or delay pesticide drops. A political party was even formed for the 2017 general election with the purpose of stopping 1080 applications, but the Ban 1080 Party was disbanded after receiving only 0.1% of the vote (3005 votes in total).[4]

This short and wildly simplified summary of 1080 in New Zealand is a great example of the pest control debate. For any pest species, such as fruit fly, possums,

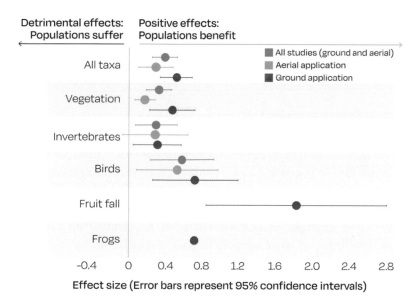

Fig. 5.2: What are the biodiversity outcomes from possum-focused pest control? This 'meta-analysis' comes from 47 studies examining ground-based control using traps and/or poison bait and aerial applications of 1080. It compares treatment areas (possum-focused pest control) and non-treatment areas (no pest control) in the response of native biota (grouped by taxa on the left side of the plot). Bars around the means denote 95% confidence intervals. Positive effect sizes indicate that pest control (treatment) areas had positive biodiversity outcomes, compared to non-treatment areas. *Adapted from Byrom et al. (2016).*[85]

rats or rabbits, a full spectrum of management options is available. We could do nothing, take no action, which is a management response. Doing nothing might even sometimes be the best option for some species. But if we do want to do something, what are the tools, and which should we use? We could use pesticides, biological control agents or even genetically modified organisms that might offer an opportunity to eradicate populations. These or any pest control methods are often imperfect and can have harmful, unintended consequences. Often different experts and different sectors of the public will disagree. There will be a wide spectrum of passionate views, more often than not unwavering and inflexible, regarding the control of any pest species.

In this chapter I will describe the triggers for pest control and provide an overview of the different options for pest population regulation. Some of these methods have been used or could be adapted for eradication purposes. Along the way, I'll try to acknowledge the many different voices, perspectives and views on pest control. How do we go about deciding if we should actively manage a pest population, and which pest control method should be used?

## The triggers for pest control

The bottom line for determining if or when to control a pest population is its population density. In triggering a pest control response, managers are attempting to limit pest abundance, now or in the future, to densities below those that would be deemed to have a significant or unacceptable impact.

For the above example of possum, stoat or rat control by DOC in New Zealand, a significant or unacceptable impact might be harm caused to native plant or animal populations by these pests. In many situations, however, we don't know what the pest abundance is that will cause unacceptable outcomes. A report from DOC stated, 'We currently do not know the intensity of pest control or maximum allowable residual pest abundance targets for many native species. For pest control to be effective it is necessary that pest populations are maintained below levels that negatively impact on native animal populations. Few examples of such maximum pest abundance targets (e.g. < 10% rat tracking, < 20% stoat tracking and < 10% possum residual trap catch over the kōkako breeding season) are known. Knowledge of such targets allows for optimisation of the timing, frequency and intensity of control to protect native species.'[5] The authors are saying that we don't know the density of possums, stoats and rats that our native flora and fauna can tolerate without experiencing population declines. The pests can breed, and populations can grow very quickly, especially when there is an abundance of food. Consequently, the management of these pest species is typically based on the time since the last pesticide application rather than directly on pest density estimates. The triggers for pest control are often based on knowledge of the reproductive rates of pest populations and their reinvasion rates.

We also know that environmental variation can swiftly and substantially change population growth rates. Mammalian pests can quickly reach plague numbers following beech (*Lophozonia* spp.) and rimu (*Dacrydium cupressinum*) 'mast seeding' events. Plants such as beech trees don't produce seed every year; instead, every 2–6 years there is a simultaneous mass-flowering and seed-production event, called masting. In mast years, entire beech forests produce huge amounts of seed. The seed provides food and promotes the abundance of seed consumers such as mice, which in turn promotes the population growth of mouse predators such as stoats.

One example cited in the DOC report was kea nesting success. With stoat control, kea (*Nestor notabilis*) nesting success was 79%. Without stoat control in a non-mast year it was 37%, but in a mast year, kea nesting success dropped to just 3% due to a subsequent plague of stoats.[5] Seed mast years are predictable, and conservation authorities can spring into action by triggering pest management to curtail pest plagues when a mast year is predicted.

Much more is known and is often refined for pest control triggers in cropping situations. Pests in cropping systems are defined by their abundance as well as the type of injury or nuisance that they cause. For many pests and crop systems, it is possible to develop an economic-injury level (EIL), which is 'the lowest population density of a pest that will cause economic damage; or the amount of pest injury which will justify the cost of control'.[6] Estimates of the EIL are typically market driven, varying between regions and seasons and with a producer's changing scale of economic values. Once an EIL is estimated, it is also possible to estimate an economic threshold or 'the density at which control measures should be determined to prevent an increasing pest population from reaching the economic-injury level'. The economic threshold is always lower than the EIL, which permits time for the application of control measures and then for these measures to take effect. A simplified equation for calculating the EIL is:

$$EIL = (C/V) \times (1/L)$$

Where $C$ = pest management costs ($ per unit measure); $V$ = the market value of product or managed resource ($ per unit measure); and $L$ = loss caused to product or managed resource, on a per pest basis (loss per unit measure per pest).[7] The optimal goal for pest managers is to maintain pest populations at an equilibrium threshold or population density well below both the economic threshold and the EIL. The introduction of new management techniques, such as a new pesticide toxic to biological control agents, can disrupt pest management programmes and drive pest densities above these EILs (Fig. 5.3). Estimates of EIL are typically used for pests that directly damage a crop by chewing or consumption. The levels are not used for pests that transmit plant or animal disease, for which their first appearance triggers an economic threshold.

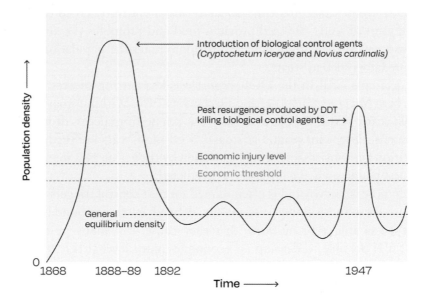

Fig. 5.3: Originating from Australia, the cottony-cushion scale was first noticed in California in 1886. Populations grew quickly, exceeding economic thresholds and economic injury levels. Densities declined after the introduction of biological control agents from Australia in 1888 but increased dramatically around 1947 due to extensive use of DDT, which killed many biological control agents. Many herbivorous pests quickly developed resistance to DDT. *Adapted from Stern et al. (1959).[6]*

A cottony-cushion scale (*Icerya purchasi*) and one of its major predators and biological control agents, a cardinal ladybird (*Novius cardinalis*). *Left: Lucarelli / Wikimedia Commons. Right: Katja Schulz / Wikimedia Commons*

Ideally, the trigger for pest control should always involve monitoring. Accurately estimating pest abundance and monitoring their densities is a critical component of most pest management programmes. An accurate assessment of pest population densities allows cost-efficiencies in spending on control options and can maximise biodiversity benefits. With pesticides as control techniques, monitoring can also enable pesticide resistance to be minimised.

## Pesticides and a history of chemical control of pest populations

The FAO and the WHO jointly define a pesticide as 'any substance or mixture of substances of chemical or biological ingredients, intended for repelling, destroying or controlling any pest, or regulating plant growth'.[8] This pesticide definition includes biopesticides. Organisations such as the United States Environmental Protection Authority label biopesticides as naturally occurring substances that control pests (biochemical pesticides), microorganisms that control pests (microbial pesticides), and pesticidal substances that are produced by plants to which humans have added genetic material (plant-incorporated protectants). Wikipedia tells us that in 2006 and 2007, the world used approximately 2.4 megatonnes of pesticides. Herbicides constituted the biggest contributor to world pesticide use at 40%, followed by insecticides (17%) and fungicides (10%). The remainder includes rodenticides, molluscicides, virucides and chemicals, such as 1080, for controlling mammalian pests.

The first recorded use of pesticide was 4500 years ago in ancient Mesopotamia, where sulphur was employed for the control of mites and insects attacking food. The Romans used sulphur in an attempt to kill ants and weeds in the 1st century BCE; the Chinese used arsenic to control insect pests in their crops and orchards in 800 CE. Over time, this developed into the use of metal compounds as pesticides, with the invention of 'Paris green' (a combination of copper acetate and arsenic trioxide) in 1814. By the 1880s, Paris green was the first globally widespread chemical pesticide. It was used to kill mosquitoes and tobacco budworm, and acquired its name when used to kill rats in Parisian sewers.

Another of the earliest widely used pesticides was DDT (Dichloro-diphenyl-trichloroethane). DDT is an organochlorine pesticide, first synthesised in 1874. Its insecticidal properties were discovered in 1939, and it was first used during World War II to combat malaria, typhus and lice among civilians and troops. Subsequently it was employed all over the world as an agricultural and household pesticide. It was heavily used for pest control on a wide variety of crops and played a substantial role in combating mosquito-borne disease. Largely because of the

Synthetic chemical pesticides have been a mainstay for pest control around the globe since the 1940s. Here, a truck sprays 'harmless to humans' DDT on a beach in Long Island, New York, 1945. *Pictorial Press Ltd / Alamy*

availability of DDT and the drug chloroquine, in 1955 the WHO launched the Global Malaria Eradication Programme with major effect. For example, the occurrence of malaria in Sri Lanka plummeted from an estimated 2.8 million cases in 1946 to just 17 cases in 1966.[9]

Problems with DDT use soon became apparent. Although 'there is no convincing evidence that DDT or its metabolite DDE [dichloro-diphenyl-dichloroethylene] increase human cancer risk', it has been associated with pre-term birth, early baby weaning, and reproductive and developmental issues in humans.[10] Rachel Carson published *Silent Spring* in 1962, in which she highlighted the effects on DDT on biodiversity.[11] DDT bioaccumulates in fat, magnifying and becoming more concentrated as you move up the trophic levels from herbivores to predators. One highly publicised effect of DDT and its breakdown products was eggshell thinning in birds such as the bald eagle. The shells would become so thin that nesting females would accidentally crush their eggs. By 1963 the global population of America's national bird fell to only 417 pairs successfully raising chicks. The species seemed doomed to extinction.[12]

*Silent Spring* received fierce condemnation from chemical companies. One spokesperson for the chemical industry commented: 'If man were to follow the

The Malaysian government spraying insecticide smoke, known as mosquito fogging, to prevent the spread of dengue fever in Batu Ferringhi on Penand Island, 2018. This pesticide and pest control method will kill many more insects than just mosquitoes. *Nalidsa / Alamy*

teachings of Miss Carson, we would return to the Dark Ages, and the insects and diseases and vermin would once again inherit the earth'.[13] Nevertheless, *Silent Spring* began a movement that ultimately led to a nationwide ban of DDT in the United States in 1972. Bald eagle populations have since begun a widespread recovery.

My favourite story about the complexities of DDT comes from the science historian Hannah Gay, who describes work by WHO in Sabah (previously North Borneo) and Sarawak, Malaysia. WHO teams were managing an indoor DDT spraying programme against mosquitoes during the 1950s.

> Malaria mosquitoes were effectively controlled, but there were problems. For example, many people living in homes with thatched roofs complained that, after the spraying, their roofs were eaten away by caterpillars. It turned out that a parasitic wasp that preyed on the caterpillars was especially vulnerable to DDT, leaving the caterpillar populations less controlled. It was also the case that many domestic cats died. They were poisoned not only because they cleaned their fur by licking it, but because they caught and ate small geckos that lived inside people's houses. The geckos preyed on insects. The cat deaths appeared to provide early evidence of DDT working its way up the food chain, and that, at certain concentrations, DDT was harmful to mammals. The death of the wasps was a lesson in the loss of biological control. So, too, was the death of the cats, as the rodent population flourished. Replacement cats had to be trucked into the most affected areas. In more remote areas, some [cats] were even parachuted in.[14]

Biodiversity is still affected by DDT, even in remote places where you wouldn't imagine it as present or a problem. Rates of DDT contamination in Adélie penguins (*Pygoscelis adeliae*) from the Western Antarctic Peninsula, for example, have been surprisingly consistent, even decades after most countries ceased using the pesticide. It is thought that DDT has been captured in frozen glaciers, from which annual meltwater is contaminating the food chain of these birds.[15] In another study, this time of humans, DDT exposure for women in the 1960s caused dramatic increases in the likelihood of obesity and early menarche in their granddaughters.[16] The harmful and serious effects of DDT exposure extended across three generations to children who haven't seen or been exposed to this pesticide.

In 2004 DDT was globally banned under the Stockholm Convention on

Persistent Organic Pollutants, which is described as 'a global treaty to protect human health and the environment from chemicals that remain intact in the environment for long periods, become widely distributed geographically, accumulate in the fatty tissue of humans and wildlife, and have harmful impacts on human health or on the environment'. The convention requires its parties to take measures to eliminate or reduce the release of persistent organic pollutants, including DDT, into the environment.

It might thus come as a surprise to many that DDT is still produced and used around the globe on a substantial scale. Even under this convention, countries can utilise DDT for 'acceptable purposes' such as the control of mosquitoes, malaria and the disease leishmaniasis. A 2012 report found that the global use of insecticides to control human disease vectors was dominated by DDT in terms of quantity applied (71% of total), although pyrethroids came first in terms of surface area treated (81% of total).[17] Eighteen countries are listed as registered users of DDT, mostly for the control of malaria and leishmaniasis, with three still able to produce the pesticide. On the bright side, a 30% decline in DDT use was observed over 2001–14, from 5388 metric tonnes down to 3772 each year.[18] India has been by far the largest producer and user of DDT, and many other countries need DDT because of mosquito resistance to other insecticides such as pyrethroids (Fig. 5.4). I think it is easy to be shocked and dismayed by the continued production and use of DDT. But I also wonder what we'd do if we were living in a low-income country, struggling with high rates of vector-borne diseases, and unable to afford alternative and more expensive chemicals or control methods. Use DDT, or let people suffer?

Some academics argue for the continued or even increased use of DDT for mosquito and malaria control. They believe the benefits of alleviating the burden of malaria outweigh the risks to human health and the environment. The DDT ban that wealthy countries insist on has been considered 'eco-colonialism' by some analysts, who argue that it 'can impoverish no less than the imperial colonialism of the past did'.[9]

## Modern-day pesticide problems: The glyphosate debate

Herbicides are the most commonly used pesticides, and glyphosate is by far the most commonly used of all the herbicides.[19] In many respects, the modern-day issues with glyphosate are typical of many pesticides and mirror the historic controversy with DDT. Glyphosate is used worldwide and is frequently considered vital by farmers. But it is claimed to have human health and environmental effects, with overuse leading to the development of resistance in weeds. The widespread use of genetically modified crops has caused an increase in glyphosate use and magnified all these problems.

Glyphosate is a broad-spectrum herbicide and crop desiccant that was discovered in 1970. An organophosphorus compound, it acts by inhibiting the enzyme system common in plants (but not in animals such as humans). It is used in almost every nation in the world to kill weeds. Its use in agriculture has been especially prominent where it is sprayed to kill annual broadleaf weeds and grasses that compete with crops. A key advantage of this chemical is that it can be used in 'no-till' farming, conserving soil. It also seems to degrade quickly.

There are two key points to note regarding glyphosate in formulations that are used by farmers. First, glyphosate is not sprayed onto crops in its pure form on its own. Chemicals called adjuvants help enhance the application, uptake and translocation of pesticides, including glyphosate. Adjuvants may be damaging to human health or the environment in their own right. The common glyphosate herbicide Roundup has the adjuvant polyoxyethylene amine (POEA). Second, pesticides such as glyphosate break down and form other chemicals. For glyphosate, the first decomposition product is aminomethyl phosphonic acid (AMPA).[20] The toxicity and metabolic pathways affected by these breakdown products may be very different than those for the initial undegraded pesticide itself. In humans, AMPA can mimic the effects of a chemical involved with the functioning of our central nervous system.[21] In DDT, a common breakdown product is dichlorodiphenyldichloroethylene (DDE), which has been described as even more toxic than DDT.

Up until the mid-1990s, glyphosate was used prior to planting crops. Then, however, came the development of genetically modified crops that were resistant to this herbicide. The development and introduction in the United States of glyphosate-resistant canola, corn, cotton and soybeans allowed the repeated application of the herbicide to control weeds all the way up to harvesting. The total area of land treated with glyphosate has increased tenfold since the introduction of these genetically modified plants. Countries outside of the United States quickly adopted this technology. Canadian farmers grew 11.6 million hectares of genetically engineered crops in 2012, with herbicide tolerance as the primary genetic modification. These glyphosate-based growing systems have been seen as effective, easy, economical, safe and simple.[22]

The repeated application of any pesticide represents a major selective pressure on a pest population. The selection is for resistant individuals. This selection works via the tiny number of individuals in a population that have naturally occurring genetic mutations that enable them to survive a pesticide and reproduce while their neighbours die. These mutant individuals might be exceedingly rare, popping up only once in millions or billions of individuals, but if tens of millions of hectares are sprayed for many decades, with each hectare plot containing hundreds or thousands of weeds, the chances of finding and selecting mutant individuals are high. There are now at least 435 unique cases of weeds becoming resistant to herbicides around the globe. Pesticide resistance is 'surging', even in New Zealand, with one survey finding resistant weeds on 54 of 87 properties.[23] Globally, at least 30 weed species have shown glyphosate resistance. The mutations in weeds that confer glyphosate resistance have been suggested to be

**Repeated pesticide applications over time**

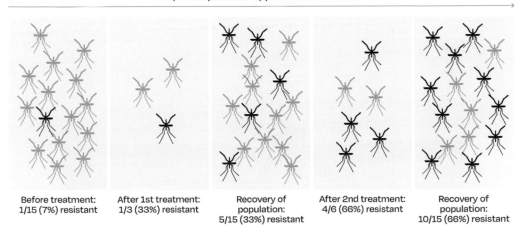

| Before treatment: | After 1st treatment: | Recovery of | After 2nd treatment: | Recovery of |
| 1/15 (7%) resistant | 1/3 (33%) resistant | population: | 4/6 (66%) resistant | population: |
| | | 5/15 (33%) resistant | | 10/15 (66%) resistant |

Fig. 5.4. Selection for resistant genotypes (shown as dark individuals) in a population when treated with a pesticide. Before treatments, mutations within the population that enable pesticide resistance are rare. After repeated treatments and selection pressure, these rare genotypes can become more abundant. With repeated selection they can comprise the majority of the population. There is often a 'cost' to being resistant: the resistant genotypes may produce fewer offspring. Resistance can be managed by alternating between pesticides with different modes of action.

the most important pest problem for global agriculture.[22] Unfortunately, farmers have often reacted to glyphosate resistance in weeds by increasing the dosage and frequency of glyphosate sprays. That reaction seems likely to increase the selection for resistant strains. Seed-producing companies have responded by producing genetically modified crop varieties with combined resistance to glyphosate and other pesticides, such as 2,4-D (2,4-Dichlorophenoxyacetic acid).[20]

With a high rate of glyphosate application over millions of hectares all over the world, we should expect to see environmental and food contamination by glyphosate, adjuvants such as POEA, and decomposition products including AMPA. And we frequently see these chemicals in soil. One study in France found 91% of water samples from streams positive for contaminants.[20] Concentrations of glyphosate and AMPA in our food vary widely, ranging from 0.1–100 milligrams per kilogram in legumes (including soybeans), cereals, vegetable oils and rice. Glyphosate and AMPA have been detected in the urine of a high proportion (30–80%) of farm animals and farmers. In Europe and the United States, residues have been found not only in the urine of farmers but also in 44–80% of the general public, including children.[20] There is a good chance that you'll eat some glyphosate or its breakdown products today.

A high clearance sprayer applies Roundup to a soybean field. *Bill Barksdale / Alamy*

The good news is that the technical grade glyphosate has an exceedingly low acute oral toxicity. The $LD_{50}$, or the dose of glyphosate required to kill half the members of a tested population after a specified test duration, ranges between 800 and 1340 milligrams per kilogram for many mammal species.[24] We scientists don't do $LD_{50}$ experiments on humans, but it is reasonable to assume that people experience a low acute toxicity to glyphosate too. Of note, however, are studies with rodents in which the Roundup adjuvant POEA has a fourfold higher acute toxicity than glyphosate.[20]

The bad news is that there is increasing evidence for negative effects of chronic exposure to formulated glyphosate and its breakdown products.[20] Studies have shown evidence of biodiversity effects. For example, in honey bees (*Apis mellifera*), exposure to glyphosate perturbs their beneficial gut microbiota, which potentially affects bee health and their effectiveness as pollinators.[25] An even bigger effect for bees may be a reduction in pollen and nectar availability via the reduced abundance of weeds (which the herbicide was applied to kill).[26] In humans, epidemiological studies have suggested correlational links between long-term glyphosate exposure and diseases including various forms of cancer, kidney damage, and developmental or mental conditions such as ADHD, autism, Alzheimer's disease and Parkinson's disease. The increase in neurological defects and conditions make a degree of sense, given exposure to the Roundup breakdown product AMPA. One study found an increase in infertility and malformation among pigs that was correlated with glyphosate concentrations in the animals' livers and kidneys. It is important to note here, however, that these were *correlational* studies, which can be confounded by other factors. *Experimental* studies have often used Roundup, which may be representative of what happens in a more realistic world that is sprayed by Roundup and not just glyphosate alone. These chronic exposure studies have found evidence of cancer, and liver and kidney damage in animals such as rats. Relatively low doses of Roundup appear to negatively affect fertility and hormone expression.

It is important to bring caution and a critical eye to the interpretation of any scientific study. Scientists are like everyone else: to a greater or lesser extent we bring our own biases and beliefs to our work. Just because a study is published in a peer-reviewed scientific journal doesn't mean it is accurate and true. This was exemplified in a study by Séralini et al., 'Long term toxicity of a Roundup herbicide and a Roundup-tolerant genetically modified maize', published in *Food and Chemical Toxicology* in 2012. The authors claimed that rats fed with genetically modified corn or with Roundup suffered tumours and organ damage. Following publication, protesters took to the streets in Brussels in their thousands to demand an overhaul of food policy in Europe. Other scientists, however, were intensely critical of the study and flooded the journal with complaints. They noted serious deficiencies in the work, including an extraordinarily low sample size, unconventional statistics that seemed to go fishing for significant effects, the fact that the type of rat used in the study was very prone to tumours, and inappropriate controls. The journal quickly retracted the study.*

In response to many of these studies largely based on Roundup, in 2015 the WHO International Agency for Research on Cancer (IARC) reclassified glyphosate as probably carcinogenic for humans. California and several countries, including Brazil, Denmark, El Salvador, France, India, Norway, Sri Lanka, Sweden and the Netherlands then proceeded to ban or restrict the use of glyphosate. New Zealand did not implement a ban and took a different approach. The Environmental Protection Authority (EPA) concluded that 'based on a weight of evidence approach, taking into account the quality and reliability of the available data, glyphosate is unlikely to be genotoxic or carcinogenic to humans and does not require classification under HSNO as a carcinogen or mutagen'.[27] The chief scientist of the EPA at the time, Dr Jacqueline Rowarth, argued, 'We need to consider everything otherwise it is not the net benefit approach' and 'we go to all sources because there is an economic implication within the use of glyphosate . . . We agree with them [the IARC] – at high exposures and dosages, cancer could occur but we don't have these high exposures and dosages in New Zealand.'[27] Several public health researchers were appalled and demanded the EPA withdraw their conclusion and reassess glyphosate use in New Zealand.[27]

The problems and debate over glyphosate are typical for many pesticides. New Zealand will continue discussing and debating 1080 and the use of neonicotinoid pesticides, as well as glyphosate. With any of these pesticides, it will always be

---

* The retraction was supported by many. As one respondent said: 'I feel that this paper was about perpetuating fear with soft statistics and conclusions that overstep the data, rather than providing sound science.' Others supported the study despite what appeared to be fundamental flaws in the work. One New Zealand scientist stated: 'It is my view that the recent study is a valuable contribution to the scientific literature, debate and process of evaluating technologies. I trust your journal to publish quality science and you have vindicated my trust.' To read other commentaries and the retraction, see sciencedirect.com/science/article/pii/S0278691512005637

possible to find evidence of some degree of harm caused by their use. Different people will tolerate different levels of risk of harm.

The extreme, precautionary approach to avoid any harm would be to stop using these chemicals altogether. This idea has been considered in Switzerland. The Swiss democracy allows a binding referendum, decided on by popular vote, to be launched with sufficient public support. In 2018 the referendum 'Save Switzerland from synthetic pesticides' was officially lodged with the Swiss government. The proposal would have entirely banned the use of synthetic pesticides in Swiss agriculture, as well as on private gardens and public land. Food produced in other countries using synthetic pesticides would have been banned from importation. The use of herbicides, fungicides, pesticides such as neonicotinoids, and weed killers such as glyphosate would cease. The referendum organisers believed that Switzerland would be able to produce enough food without these chemicals. The 10-year lead-in from the legislation passing to it being enforced would have allowed time for new alternative biopesticides or pest control methods to be developed. The Swiss government was not supportive and in June 2021 the proposal was overwhelmingly rejected. The director of the Swiss Farmers' Union, Martin Rufer, said a total ban on synthetic pesticides would have been 'unrealistic'. The proposal would have had major consequences for the agricultural sector and the country. Swiss farmers want to use fewer pesticides but there are not enough viable alternatives to stop completely.[28] The Swiss story is very relevant, however, in reflecting the modern-day level of concern that people have about pesticide use.

## Trends in pesticide use and 'softer' pesticides

We have good data on global trends in pesticide use from the FAO.[29] The global use of herbicides in 2018 was slightly more than double that of 1990 levels (Fig. 5.5). There was a sharp rise in herbicide use for the period from 1990 to 2012, after which it appeared to stabilise. The use of fungicides and bactericides has been slowly increasing. Insecticide use appears to be relatively consistent, although there are at least two worrying issues or trends. First, weight-based measures may not necessarily be informative. Researchers have found that the toxicity of insecticides to pollinators and aquatic invertebrates has increased considerably over the last 25 years, largely driven by the development of neonicotinoids and pyrethroids that are highly toxic at very low amounts.[30] Second, the patterns of insecticide use are thought to be changing. India, for example, was a strong adopter of genetically modified *Bt* cotton, a variety of cotton that has been genetically modified to produce a bioinsecticide, taken from the bacterium *Bacillus thuringiensis*, to combat insect pests. India's adoption of *Bt* cotton enabled a substantial reduction in insecticide use. But that reduction is now reversing as pest resistance is becoming widespread in India. As Kranthi and Stone concluded, 'With *Bt* resistance in another pest and surging populations of non-target pests, farmers now spend more on pesticides today than before the introduction of *Bt*. Indications are that the situation will continue to deteriorate.'[31]

There are, fortunately, pesticides that can be less environmentally harmful. Insect growth regulators stop or slow insect growth by interfering with their metabolism or development. They may stop a pest's hormones from acting properly (such as juvenile hormone mimics that stop pests from developing

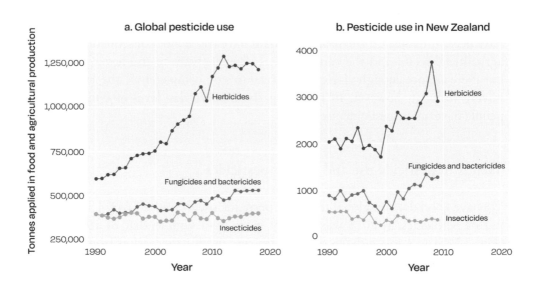

Fig. 5.5. Trends in pesticide use globally and in New Zealand. The Pesticides Use database includes data on the use of major pesticide groups. Its data report the quantities (in tonnes of active ingredients) of pesticides used in or sold to the agricultural sector for crops and seeds. Data for New Zealand appears reliable up until 2009, after which the same value for the 2009 is repeated. *Food and Agriculture Organization's Pesticides Use database*[29]

properly). Biopesticides are another 'softer' form of pest control, defined as mass-produced agents manufactured from living microorganisms or natural products that are sold for pest control. Some biopesticides target vertebrates and plants, but *Bt* is the most common and widely used microbial biopesticide.[32] The bacterium naturally produces a protein crystal (the *Bt* d-endotoxin) during bacterial spore formation in the gut of insects, causing gut cell lysis and the death of susceptible insects. *Bt* can be applied as a biopesticide or the genes for crystal production can be incorporated into genetically modified plants, as with the cotton example in the previous paragraph. Other mass-produced microbial insecticides are based on entomopathogenic fungi and viruses.

Several virus-based biopesticides have good target pest specificity. Those based on biochemicals include the widely used Spinosad, which is an insecticide based on chemicals isolated from the bacterial species *Saccharopolyspora spinosa*. Spinosad was discovered in 1985 in isolates from crushed sugarcane in the Virgin Islands, which affected the insect nervous system but with extremely low toxicity to vertebrates. It was used in New Zealand in the eradication programme for the large white butterfly (*Pieris brassicae*) as discussed in Chapter 4.[33] Note that biopesticides including Spinosad are not perfect: they can still be highly toxic, exhibiting lethal and sublethal effects for non-target and beneficial insects including bees.[34]

The 'holy grail' for a pesticide or biopesticide would be something that offers control for only the target species, leaving other organisms completely unaffected. The nearest technology we currently have that would achieve this goal is the use of dsRNA (double-stranded ribonucleic acid) for gene silencing. The gene-silencing approach is based on the application of dsRNA, which induces sequence-specific messenger-RNA degradation that halts the production of essential proteins or biochemicals within pest species. Pathogen-specific dsRNA has been shown to be effective for virus control in plants.[35] Gene silencing is seen as the next generation of pesticides and has 'enormous potential for applied entomology'.[36] To date, there has been only one proof-of-concept study, using gene silencing to control sea lamprey pests.[37] Scientists have speculated, however, that dsRNA could be used for vertebrate pest management in a variety of ways, including fertility control, invasive species eradication and pest species control to protect human health and agriculture.[38]

## Pest management using semiochemicals such as pheromones

Semiochemicals are chemical communication signals that are released into the environment by living organisms. They transmit information between individuals and are classed as either pheromones or allelomones. Pheromones are molecules that act as signals to produce behavioural or endocrinal reactions in members of the same species. They play several different roles, such as in sex attraction, signalling between parents and offspring, social organisation, and the priming of the reproductive cycles. Allelomones will modify the behaviour of individuals belonging to a different species.[39]

Pheromones are used for pest management in four main methods. First, population monitoring uses pheromones in baited traps, such as with the New Zealand monitoring network for fruit flies that uses female sex pheromones and attracts only males. Second, mass trapping uses a large number of trapping devices that can catch multiple individuals. The primary objective of mass trapping is to capture enough pests in the treated area before they damage crops or livestock, or reproduce. Mass trapping using pheromones has proven complementary and necessary for the management of some insect pests such as the boll weevil (*Anthonomus grandis*) on cotton plants in the United States.[40] Third, pheromones can be used in combination with insecticides (lure-and-kill), or with insect pathogens (lure-and-infect). The combination of pesticides and synthetic sand-fly pheromones effectively controls sand flies (*Lutzomyia longipalpis*) that are the vector for leishmaniasis in Brazil.[41] Finally, they can be used in mating disruption by permeating an area with pheromones. For example, an aerial application of sex pheromones for the sensory impairment and masking of females from males has been proposed as a potential eradication tool for the light brown apple moth (*Epiphyas postvittana*) in California.[42]

Pheromones can play a major role in the life history of mammalian pest species. Consequently, several potential applications have been suggested for pheromones in mammalian pest management. Synthetic pheromones or allelomones could play a major role in population monitoring. Conventional monitoring techniques can fail to detect some individuals, such as a known lone Norway rat (*Rattus norvegicus*) on an island,[43] so attractants to improve monitoring efficacy would be welcomed.

Commercially produced mammalian pheromones have been used in pest management and eradication programmes. The commercial cat pheromone

Feliway (produced by Virbac) was employed in an eradication programme for cats (*Felis catus*) on the subantarctic Macquarie Island, along with a range of allelomones including a commercial bobcat lure (produced by Russ Carman) and cat repellent Skunkshot (produced by Victoria Link Ltd).[44]

New technologies will be needed to achieve New Zealand's goal of becoming predator free by 2050. Pheromones or allelomones may be developed as useful tools, but they have challenges. For example, the same chemical may attract one mammalian pest but deter another.[39]

## Biological control using natural enemies

Biological control is the human use of specific living organisms (including viruses) to suppress population densities or reduce the impact of pest species. The ideal is to drive pest populations down to an equilibrium level well below where they cause economic injury or environmental harm. Biological control (or biocontrol) will generally not result in the eradication of pests. Four types of biological control are recognised by pest managers.

*Classical biological control* is relevant to the 'enemy release hypothesis' for biological invasions, where pests have been introduced from a different region but have left behind their natural enemies.[45] It involves searching for and importing natural enemies (predators, parasitoids or pathogens) from the home range to limit the density of the pest species. The introduction of the cardinal ladybird (*Novius cardinalis*) from Australia in 1888/1889 to control cottony-cushion scale (*Icerya purchasi*) in California is one example (Fig. 5.3). The unfortunate introduction of stoats to New Zealand in 1884 for the control of rabbits and hares is another.

*Neoclassical biological control*, similar to classical biological control, is a non-native natural enemy imported to control pest densities. However, in neoclassical biological control, the natural enemy has not co-evolved with the pest. It is sometimes termed 'new-association biocontrol'.[46] Typically, the natural enemy is a generalist predator that attacks a wide variety of prey species. The harlequin ladybird (*Harmonia axyridis*) is an example. Native to Asia, this predator has been intentionally introduced to many countries as a generalist biocontrol agent. In several areas it has become the most abundant ladybeetle, sometimes suppressing pests, but it has also been associated with the decline of native ladybeetles via intraguild predation and competition.[47] This predator has an obligate parasitic microsporidia that does not harm the harlequin ladybird but is a lethal pathogen for other native ladybirds.[48] Neoclassical biological control is seen by many to pose an extreme risk, as the generalist predators may become invasive pests.[49]

*Augmentation of natural enemies* involves the supplementation of natural enemies in communities. Pest managers can add natural enemies when their populations may be low or are not permanently established in an area. This supplementation may be in the form of periodic releases, or perhaps inoculative releases of individuals. It is extremely popular in Latin America and Asia, where it is seen as healthier for humans, inexpensive, and successfully saves on costs and increases productivity.[50] Another form of augmentation is inundation or inundative releases, wherein large and overwhelming numbers of natural enemies are released. 'Bioherbicides' are a form of inundative release. For example, SolviNix is a commercially available foliar bioherbicide that is based on the Tobacco mild green mosaic tobamovirus. The plant virus is mass-produced and is sprayed to control a weed called tropical soda apple (*Solanum viarum*) in the United States.[51]

*Conservation biological control* seeks to protect or enhance the activities of natural enemies that are already present in an environment. Sometimes this approach involves reducing pesticide use to preserve populations of existing natural enemies. Environmental manipulation can also conserve natural enemy abundance or modify interspecific interactions. Conservation biological control has been heavily promoted in agricultural landscapes with the planting of alternative crops or refuges that can increase the availability of food (for example, nectar, pollen) and habitat for predators and parasitoids.[52] In Asia, plants that provide nectar on the borders of rice fields were found to promote biological pest control, which ultimately led to a trophic cascade that increased grain yields.[53] Pesticide use in these Asian rice fields was also substantially lower. The use of conservation biological control has even been extended to the management of human pathogens. Lyme disease, for example, is a caused by a bacterial pathogen (*Borrelia* spp.) spread by ticks. The disease is a zoonosis, spread to humans from a natural reservoir of ticks feeding on birds, small mammals and even deer. Populations of ticks are much lower in the absence of deer. Consequently, scientists have hypothesised that conserving an abundant population of predators of deer will alter the host abundance of ticks, the ticks themselves and the disease prevalence in humans.[54]

One of the earliest successful examples of classical biological control was the introduction of the cactus moth (*Cactoblastis cactorum*) from South America to control prickly pear cactus (mostly *Opuntia stricta*) in Australia. The prickly pear cactus was introduced by European settlers to Australia as fences for agriculture and an attempt to start a clothing dye industry. The cactus quickly spread out of control. By the 1920s an estimated 24 million hectares of Australia were infested with prickly pear. Approximately half of this land was deemed so badly infested that it was 'useless from a productive viewpoint'.[55] Alan Dodd, the Australian entomologist responsible for the moth's introduction in 1926, described the spectacular outcome:

> In August 1932, 90% of the [prickly] pear had collapsed. The change in exactly two years was extraordinary . . . The prickly pear territory has been transformed as though by magic from a wilderness to a scene of prosperous endeavour . . . the most optimistic scientific opinion could not have foreseen the extent and completeness of the destruction. The spectacle of mile after mile of heavy [prickly] pear growth collapsing en masse and disappearing in the short space of a few years did not appear to fall within the bounds of possibility.[55]

Australian explorer and author Michael Terry standing in a patch of prickly pear, Bingara, New South Wales, 1921. Approximately 24 million hectares in Australia were infested with this cactus. *National Library of Australia*

The moth didn't work alone. Its devastating effects on the cactus were partly based on the feeding wounds providing access points for plant pathogens that contributed to the cactus deaths.[56] You can still find monuments and town halls commemorating the moth (though not the plant pathogens) throughout Queensland.

Heroes can become villains, even in pest management. The cactus moth is known as a generalist species with a broad diet. It has been introduced widely around the world for the biological control of prickly pear cactus species. In 1957 a decision was made to introduce it to control native *Opuntia* species in the Caribbean. It seems likely that the moth was then shipped to Florida via the plant nursery trade or by cactus collectors, with frequent interceptions at ports of entry during the 1980s. There are very rare species of cacti of conservation importance in Florida. There is also subsistence cactus farming in Mexico, based on by-products including juices, jams, confectioneries, pharmaceuticals and cosmetics. Farmers and industries relying on prickly pear cactus are now under major threat from the moth. The farmers consider it a pest in need of management. The introduction of parasitoids to attack the moth has been proposed. Eradication is being considered.[55]

Larvae of the cactus moth feeding on an *Opuntia* sp.
*Ignacio Baez / USDA*

Research on and the application of biological control flourished after Rachel Carson's *Silent Spring*, which alerted the world to the problems with pesticides. But, just as with pesticides, well-intentioned attempts at pest management using biological control can also cause substantial problems and have led to extinctions, such as with the introduction of stoats and ferrets in New Zealand. One meta-analysis estimated that 11% of all introductions for biological control purposes have likely had serious consequences for populations of non-target species.[57] Despite these issues, estimated cost–benefit ratios for classical biological control are highly favourable (1:250). For augmentative control, the cost–benefit ratios are similar to those of insecticides (between 1:2 and 1:5). Both classical and augmentative biological control have much lower development costs. It has been estimated that classical biological control is being applied to 10% of all land under culture around the globe.[58]

How do we select successful biological control agents with few non-target effects? Host-range testing is certainly key to understanding the range of species likely to be directly affected by a biological control agent, but can be difficult as it is reliant on species being able to be reared in laboratory or quarantine conditions, as discussed in the Chapter 3 example of the Japanese honeysuckle (*Lonicera japonica*) and the Honshu white admiral butterfly (*Limenitis glorifica*) in New Zealand. Difficulties rearing the butterfly precluded more substantive host-range testing,[59] which was criticised at the time. Developing a more 'natural' laboratory cage bioassay for ecological host-range testing is a priority for biological control practitioners.[57] The Hazardous Substances and New Organisms (HSNO) Act 1996 also requires evidence that the proposed biological control agent will not cause any significant deterioration of natural habitats, or have any significant adverse effects on human health and safety, or cause any significant adverse effect to New Zealand's inherent genetic diversity. In addition, it cannot cause disease; be parasitic; or become a vector for human, animal, or plant disease – unless the purpose of the importation or release is to import or release an organism to cause disease, be a parasite, or be a vector for disease (as is often the case with biological control agents).[60]

Biologically and ecologically, successful biological control agents share several common characteristics, which will be useful for future biological control agent selection.[58] The control agent should be able to tolerate and grow in all the climatic conditions in the pest's range. Successful biocontrol agents commonly show a strong searching ability to locate prey and an ability to disperse or move to prey populations. They typically have higher potential rates of increase than their prey. (A high rate of increase might be achieved via a faster developmental time, more generations per year, a greater fecundity or a greater predation rate than the reproductive ability of the pest species.) The biological control agents should be able to survive at low pest densities, and the target pest should be its strongly preferred prey. In some cases, the use of an alternative prey species might be useful during periods when the pest occurs at a low density. It is important for the life cycles of the pest and natural enemy to be synchronous.

## Integrated pest management

There are often substantial problems with using pesticides in agroecosystems. Pest populations can quickly gain pesticide resistance, while their natural enemies remain highly susceptible to these chemicals. Pest populations will then rebound quickly to even higher densities after pesticide use (termed 'pest resurgence'). Non-target organisms, including biological control agents and pollinators, can be killed, and this can induce secondary pest outbreaks. These occur when a pesticide kills the natural enemies of a species that has otherwise not been a problem, resulting in that species increasing in density to become a pest.

These problems have led to the development of Integrated Pest Management (IPM) approaches. The textbook definition of IPM is 'a pest management strategy that integrates the use of multiple suppressive tactics, often involving biological control, for optimising the control of pests in an economically and ecologically sound manner, by taking into account negative and positive impacts of pest control on producers, society and the environment'.[61] IPM seeks to prevent pest problems, identify and preserve beneficial organisms such as pests' natural enemies, and monitor pests and pest densities. It uses all tools available for pest control including natural enemies and target-specific pesticides, but also cultural pest controls (such as crop rotation) and physical control tools (such as rodent trapping).

One success story of IPM is in New Zealand's apple industry. Apple orchards host a wide variety of pests, such as the codling moth (*Cydia pomonella*) and the woolly apple aphid (*Eriosoma lanigerum*). Historically, broad-spectrum and highly toxic chemicals, including the organochlorine insecticide DDD, were used to control pests in this system. Several pests developed pesticide resistance, eliminating organophosphate insecticide use by 2001. Organochlorine insecticide and other insecticides were replaced with pest monitoring systems, threshold-based selective insecticides and biological control. Recent demands for ultra-low residues of pesticides on fruit have led growers to adopt the use of pheromones for mating disruption and the use of biological insecticides. These IPM methods have substantially reduced pest problems and have contributed to major reductions in the use of insecticides. The fruit production system now in place is considered environmentally responsible and enables access to global markets.[62]

The uptake of IPM for vertebrate pest control has been slower. Nevertheless, several projects have demonstrated how IPM principles can be used for vertebrate predators controlling vertebrate pests. In Napa Valley, California, vineyard owners installed nest boxes to attract American barn owls (*Tyto furcata*) to their properties as rodent predators. Video recordings show that each owl family removed an average of 1001 rodents each nesting cycle.[63] The careful use of rodenticides to ensure these predators were not subject to secondary poisoning,

An American barn owl (*Tyto furcata*). *Pally / Alamy*

along with the nest boxes, facilitated effective and economically beneficial pest management. The effects of these natural enemies may not be limited solely to direct predation. The presence of predators in an environment can create a 'landscape of fear' for pest rodents, deterring their foraging and effects on an ecosystem.[64]

## Genetic techniques for pest control

Genetic control methods rely on pest reproduction. By mating, the pests transmit a genetic element or modification to their offspring. This genetic element or modification then induces a feature or trait in the population that biodiversity managers might desire, such as lower rates of reproduction or lower resistance to pathogens. Genetic control technologies broadly fall into two types, based on either suppression or replacement of the pest population.[65] Suppression involves the use of genetic technologies to limit local populations, which are themselves self-limiting, eventually becoming extinct. Replacement is when the pest population is replaced by a benign or harmless form, or alternatively with a trait that is self-sustaining and driven through the pest population to reduce the harm caused by the pest.

Like biological control, there are different types of genetic control methods, which have or are being used to control pathogens, weeds, insects and other pests. Much of the work in this area has involved the use of genetic control for insect pests.

*The sterile insect technique (SIT).* In 1955 a key breakthrough was made in the development of a species-specific pest *suppression* method. The SIT releases huge numbers of mass-reared male insects that have been sterilised, typically by irradiation. The sterile males mate with females, which then produce only infertile eggs. Repeated releases of sterilised males are needed to eventually flood and overwhelm existing populations. One of the most famous uses of SIT was to eradicate the screw-worm fly (*Cochliomyia hominivorax*) from North and Central America. Screw-worm larvae can be a major pest of cattle and can even parasitise humans. They eat animal flesh after invading open wounds. The North and Central American eradication of screw-worm is estimated to have benefited the livestock industries by more than US$1.5 billion per year, which is extremely favourable compared to the cost of the entire eradication investment over more than 50 years: approximately US$1 billion.[66] Similarly, Mexico protects its vegetable and fruit export markets, which have a value of over US$3 billion each year, through an annual investment in SIT of around US$25 million. The key problem with the SIT technique, however, is that the sterile males are often less competitive than wild males. These males are mass-reared and adapted to a factory environment rather than natural conditions and are also subject to irradiation that significantly reduces their fitness.[65] An approach called the Trojan Female Technique, or TFT, is a new twist on the sterile-insect or sterile-male approach. It has been proposed that sustained population control could be achieved through the steady release of large numbers of 'Trojan females' that carry mitochondrial DNA (mtDNA) mutations causing a reduction in male, but not female, fertility.[67]

*Release of insects carrying a dominant lethal (RIDL).* The RIDL method is another pest *suppression* approach that is similar to the SIT but can overcome issues associated with fitness reductions due to irradiation. It involves a genetic modification. For example, female-specific RIDL systems involve a genetic transformation that is lethal for any female juveniles. This lethality is suppressed in mass-rearing conditions by adding the antibiotic tetracycline. Individuals can be mass-produced in the laboratory, but after release and in the absence of the antibiotic, all female offspring die. The released males and their offspring survive, passing on the genes involved in female mortality but maintaining the genetic manipulation through successive generations. This approach has been tested with several insect pests, including the Mediterranean fruit fly (*Ceratitis capitata*), diamondback moth (*Plutella xylostella*) and pink bollworm (*Pectinophora gossypiella*) (Fig. 5.6).[65] Only pest species that are able to be mass-produced are suitable for RIDL control approaches, because a lot of individuals are needed. It is currently being used on a wide scale in Florida for the control of the yellow fever mosquito (*Aedes aegypti*) and the human diseases that it spreads (see 'Case study 2: Human disease control using genetically modified mosquitoes', page 203). The RIDL approach is a commonly applied genetic modification technique for insect pest control and is being used in many countries.

*The sterile-male incompatible insect technique (IIT).* The IIT method is another population *suppression* approach, but it uses a bacterial parasite. Infections with bacteria such as *Wolbachia* have dire consequences on the outcome of mating between insects, called cytoplasmic incompatibility. We don't know exactly how it happens, but we do know that female sterility results when *Wolbachia*-infected

USDA, FDA, EPA

**USA**
Diamondback moth
and pink bollworm
field pilots
2006-2017

**UK**
Global R&D HQ

**USA**
OX5034 field pilots
2021-2022

**CAYMAN**
OX513A field pilots
2010-2018

National Office
Sanitary Secur
of Food Produ

National
Conservation
Council

**MEXICO**
Mosquito
contained pilots
2010

**PANAMA**
OX513A
contained pilots
2014

**BRAZIL**
Fall armyworm
field pilots
2019 - 2022

National
Biosafety
Committee

**BRAZIL**
OX513A
commercial
biosafety approval
**2014**

**BRAZIL**
OX5034
commercial
biosafety approval
2020

b

CTNBio

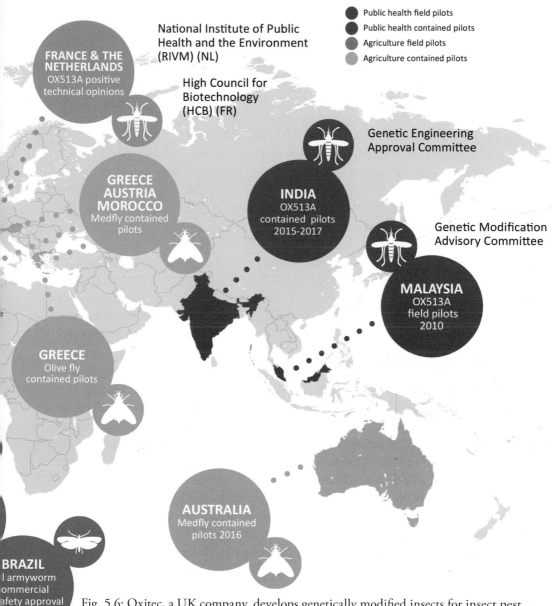

**Public health field pilots**
**Public health contained pilots**
**Agriculture field pilots**
**Agriculture contained pilots**

National Institute of Public Health and the Environment (RIVM) (NL)

High Council for Biotechnology (HCB) (FR)

Genetic Engineering Approval Committee

Genetic Modification Advisory Committee

**FRANCE & THE NETHERLANDS**
OX513A positive technical opinions

**GREECE AUSTRIA MOROCCO**
Medfly contained pilots

**INDIA**
OX513A contained pilots 2015-2017

**MALAYSIA**
OX513A field pilots 2010

**GREECE**
Olive fly contained pilots

**AUSTRALIA**
Medfly contained pilots 2016

**BRAZIL**
l armyworm ommercial afety approval 2021

Fig. 5.6: Oxitec, a UK company, develops genetically modified insects for insect pest control. The company believes that genetic methods of population control are more effective and more environmentally friendly than insecticides. Their programmes involve releasing genetically modified insects – mosquitoes, fruit flies and the diamondback moth (*Plutella xylostella*) – into several countries. They are developing approaches for the fall armyworm (*Spodoptera frugiperda*) and other pests. *Oxitec*

males mate with uninfected females. Consequently, managers might be able to suppress pest numbers if they release large numbers of *Wolbachia*-infected males into populations of non-infected females. This approach has also been used for mosquito control. Zheng et al. (2019) mass-released millions of factory-reared adult male mosquitoes in China over a two-year period and enabled near-elimination of wild-type mosquito populations that vectored human pathogens such as the dengue and Zika viruses.[68] A substantial drop in the mosquito biting rate was observed, as well as a surge in public support for the approach.

The key issue for the IIT, or RIDL or any other suppression approach, is that reinvasion can and is likely to occur. Populations of pests may be driven down for many weeks, months or years, but eventually the pest is likely to reinvade the control area. This is where *replacement* techniques may be more useful.

*Techniques for population replacement: Wolbachia-based drives in insects.* A pest control technique involving *replacement* requires some mechanism to drive genes through a population. It is essential for the desired genotype to increase in frequency to become highly dominant or abundant, in a way that vastly exceeds what would occur under normal Mendelian genetics. Several replacement approaches can be used; I'll discuss two common techniques here. The first is using *Wolbachia*-infected mosquitoes. *Wolbachia*-infected female mosquitoes have a fitness advantage over uninfected females, because females that don't have this bacteria become sterile after mating with an infected male.

*Wolbachia* can also have a major effect on viruses in mosquitoes. For example, a mosquito infected with the *w*Mel strain of *Wolbachia* can block transmission of dengue virus.[69] This approach is sometimes referred to as paratransgenesis and involves the mass-rearing and release of the *Wolbachia*-infected mosquitoes. It has now been used in several countries. In one Indonesian city after this replacement approach was used, dengue cases fell by 77%.[70] Dengue cases requiring hospitalisation were lower by 86.2% in treated zones. The study authors were overwhelmed with the outcome: 'the intervention is self-sustaining and resilient'.[71] This approach was also considered equitable because the mosquitoes do not discriminate between affluent and poor neighbourhoods. They can also be deployed cheaply.

*Techniques for population replacement: CRISPR-based gene drives.* For many animals and pest species, the chance of inheriting a gene is 50% under normal Mendelian inheritance. This outcome is because organisms typically carry two chromosomes with different genetics on each that are inherited from each parent.

A CRISPR (Clustered Regularly Interspaced Short Palindromic Repeats) system is a genome-editing mechanism that overcomes Mendelian inheritance by copying itself onto both chromosomes (Fig. 5.7). CRISPR systems are naturally common in bacteria as a defence mechanism against viruses but have been borrowed by scientists for gene editing. An organism with a CRISPR modification has the desired gene on both chromosomes and in all offspring. The highly targeted and 'programmable' modification can be driven through the population as a method of population replacement.

As of 2021, there have been no field releases of CRISPR gene drives, although several have been proposed and modelled, such as the Y-shredder or Y-CHOPE (Y-Chromosome deletion using Orthogonal Programmable Endonucleases) strategy. This strategy uses the CRISPR technology to incorporate an enzyme that

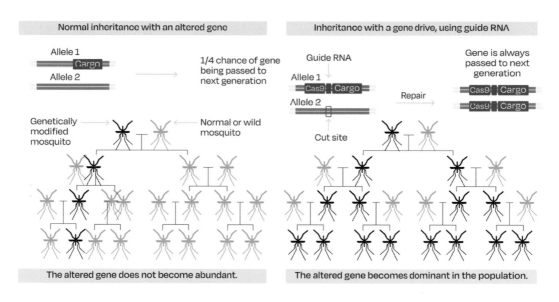

Fig 5.7. Gene drives propagate a particular gene or cargo of genes throughout a population by altering the probability that a specific allele will be transmitted to offspring. Left: Under normal inheritance, an altered gene or cargo of genes would not propagate well. It may become extinct, especially if negative fitness effects were associated with the cargo. Right: An example of a gene drive using the CRISPR-Cas9 technology harnessed from bacteria. The CRISPR-Cas9 edits genes or alleles by cutting DNA, guided by a short segment of guide RNA (gRNA). Natural DNA repair processes then repair the cut. This results in all offspring inheriting the modified gene, which then sweeps through the population over following generations.

destroys the Y-chromosome in pests such as rodents, resulting in the production of only female offspring. Conceptually, over successive generations the number of males and mating partners declines, and the population will eventually go extinct. A similar approach with an X-shredder was proposed to control mice plagues in Australia that occurred in 2021.[72] Support for this genetic control proposal was immediate, even from animal rights groups. People for the Ethical Treatment of Animals (PETA) Australia welcomed the X-shredder approach as a major advance over the use of pesticides: 'We're pleased to see the government is finally using science to tackle this problem in a more ethical and eco-friendly manner.'[73]

Many of these genetic control approaches are untested new developments, resulting in justifiable concern and debate. For example, three key issues stand out in the discussion on whether CRISPR-based gene drives represent a 'biocontrol silver bullet or global conservation threat:' (1) the importance of understanding target specificity, (2) the implications of population connectivity and (3) the need to carefully consider unintended cascades for community dynamics.[74] These are valid concerns. In our work with invasive wasps (*Vespula* spp.) in New Zealand, we have identified highly specific genetic targets for a gene drive and, with simulation modelling, have shown population regulation is possible.[75] As these species are not native to New Zealand, the impact on community dynamics would largely be beneficial, especially for native bird populations that compete with wasps for resources in native forests. An analysis of likely community effects after wasp eradication has been performed.[76] But a key concern is if wasps were to make it back to their native range, where they do provide an array of ecosystem services. Perhaps solutions could be implemented to avoid wasp movement or the effects of a gene drive in the native wasp range. Similar concerns should be voiced over the genetic modification of mice to control plagues in Australia.

Other modern genetic manipulation approaches are much better tested and have provided effective, long-term pest control. One of the earliest was genetically transforming the transgenic papaya in Hawai'i to resist the papaya ringspot virus.[77] This virus was resulting in a massive production drop from 140,000 to 5600 kilograms per hectare per year.[78] No effective control options were available, nor were there any varieties of papaya that showed resistance to the virus. One idea, however, was to genetically modify the papaya by inserting a small piece of the viral RNA (ribonucleic acid) into the genome of the papaya plants. A similar approach had been used to make tobacco plants resistant to a plant virus. Although the resistance mechanism was not yet known, it also seemed like a potential opportunity for papaya. At that time, the methods of genetic transformation were crude. 'Gene guns' took gunpowder from a .22 calibre bullet and were used to propel DNA-coated tungsten balls into plant cells. Shots would be fired hundreds or thousands of times into plant cells until, if you were lucky, the section of DNA you wanted would be encoded somewhere on the genome of the plant. The gene guns and genetic transformation was successful in papaya. We now know the plants produce a small section of the virus's genetic code in the form of RNA, which causes the plant cells to bind and 'silence' the production of similar (viral) RNA. Nearly three decades after its introduction, this genetic modification and gene-silencing approach remains an effective control for the papaya ringspot virus. If you've been to Hawai'i and eaten papaya, it is highly likely to have been this genetically modified variety. It is now approved for export to and consumption in countries such as Japan.[78]

In 2018 there were over 190 million hectares of genetically modified crops grown in 29 different countries around the world.[79] Five countries (the United States, Brazil, Argentina, Canada and India) planted 91% of the genetically modified crop area in 2019. In addition, 42 countries import produce from these crops for food, feed and processing. Soybeans, maize, cotton and canola are the most widely grown genetically modified crops, but others include alfalfa, sugarcane and potatoes. Approximately half of the 190 million hectares is occupied by soybeans that have been genetically modified to be resistant to the herbicide Roundup, and/or for the expression of a bacterial protein (the Cry1Ac protein from *Bacillus thuringiensis* or *Bt*) as a biopesticide or a modification to enhance the preservation of the oil derived from the beans. The creation of varieties with multiple genetic modifications is referred to as 'stacking'. Maize is the second-most widely grown genetically modified crop. Like soybeans, maize varieties have been modified to be resistant to herbicides, to have single or multiple genes for insect-resistance, to be virus- or drought-resistant, or to have increased nutritional value (see 'Case study 3: Genetically modified maize', page 205).

While there are environmental issues associated with the excessive use of herbicides such as Roundup in cropping systems where plants have been genetically modified to resist this herbicide, there is substantial evidence that the careful use of many genetically modified plants can be environmentally beneficial. For example, one study in China reported the effects of genetically modified cotton that was engineered to produce insecticidal toxins from *Bacillus thuringiensis* (*Bt*) and so to resist the pest cotton bollworm (*Helicoverpa armigera*).

This variety has been widely planted in Asia. Their analysis reported that the pest population dynamics from 1992 to 2007 in China

> indicated that a marked decrease in regional outbreaks of this pest in multiple crops was associated with the planting of *Bt* cotton. The study area included six provinces in northern China with an annual total of 3 million hectares of cotton and 22 million hectares of other crops (corn, peanuts, soybeans and vegetables) grown by more than 10 million resource-poor farmers. [Their] data suggest that *Bt* cotton not only controls *H. armigera* on transgenic cotton designed to resist this pest but also may reduce its presence on other host crops and may decrease the need for insecticide sprays in general.[80]

No genetically modified commercial crops are currently grown in New Zealand. Our policy on GMOs stems from a Royal Commission on Genetic Modification that was held in 2001, with subsequent amendments to the HSNO Act 1996. Given advances over the following years, an expert panel set up by Royal Society Te Apārangi in 2019 called for an overhaul of gene-technology regulations and wide public discussion. Māori will be essential in future decision-making processes regarding gene-editing technologies in New Zealand, especially with the concepts of whakapapa (genealogy), mauri (life essence) and kaitiakitanga (guardianship). Historically, most gene-editing technologies have been seen as inconsistent with Māori values. However, the relationship between gene editing and Māori values is not always negative, with surveys suggesting that mauri and kaitiakitanga might actually be enhanced through the use of gene-editing technologies.[81]

Tame Malcolm, the operations manager at Te Tira Whakamātaki, a not-for-profit Māori biodiversity network, believes that new technologies like gene editing may be key to achieving a predator-free New Zealand. 'If they change the possum's DNA, so that the possum only produces male or female [offspring] and makes it dominant, every generation after that will be female, let's say. If you give it enough cycles, enough generations, you're left with just females and then it breeds itself out. People say stuff like that is against our kawa [protocol and etiquette], against whakapapa. Rangi Matamua [an academic at the University of Waikato] makes a good point, he says if they believe in whakapapa, surely they must believe in Māui and Māui was able to change his DNA to turn into a kererū, tuna . . . so where does the conversation go after that?'[82]

## Summary

One of my goals in this chapter was to convey the difficulties of pest control. There is risk associated with *any* form of pest management, even when that management is to do nothing.

These challenges are well represented in pesticides, which are still the primary way in which we control pests. I'm sure that all farmers or biodiversity managers would much prefer to spend their money in other ways, rather than on spreading synthetic chemicals over landscapes. Their decisions to use pesticides and the ramifications of this use is complex and involves trade-offs. We know DDT has residues that linger, bioaccumulate and damage biodiversity in every corner of the world. But how do we balance these effects against the lives of people suffering from malaria in developing nations? The issues with glyphosate are similar and different. The adjuvants used in the application of glyphosate, such as in Roundup, and the breakdown products appear much more damaging than glyphosate itself. Its links to cancer, however, are hotly debated: scientists cannot agree among themselves about the best practice for managing this herbicide.[83] There is some level of cancer risk associated with glyphosate, but what is the most accurate assessment of that risk? What hope does the general public have of reaching a consensus when different scientists, trained for decades in experimental design and statistics, look at the same study only to come to vastly different conclusions? Similar debates rage on about other pesticides, such as neonicotinoids, and their effects on non-target organisms, including bees.[84]

Our knowledge on many other pesticides does offer a clearer picture and management pathway. The use of 1080 in New Zealand is one example of a pesticide that is currently needed and has a relatively low level of risk. A high majority of scientists support the continued use of 1080. While everyone would prefer to spend the hundreds of millions of dollars that are used for the production and application of 1080 differently, I believe that we need to keep using this toxin until better pest control approaches are developed.

One of those better approaches may be using genetic technologies, although the jury is still out as to whether those approaches will always work and be socially acceptable.

## Case study 1: Evaluating the use of 1080

The Parliamentary Commissioner for the Environment's report, *Evaluating the use of 1080: Predators, poisons and silent forests*, presented a framework for evaluating 1080 and its alternatives. This report focused on the use of aerial drops of 1080 for the control of possums, rats and stoats in New Zealand.[1] It is a useful framework, providing a set of nine questions that could be adapted to evaluate other pesticides or pest control methods. The questions, with a summary of answers presented for 1080, were:

1. *Can the method decrease populations of possums, rats and stoats?* Kill rates from using 1080 were generally 75–100% for possums and close to 100% for rats. There is less information on kill rates for stoats, as this occurs through secondary poisoning after stoats feed on the carcasses of rats or possums killed by 1080. But the limited information on stoat mortality suggests that kill rates are high. In summary, 96% of monitored aerial 1080 applications met pest reduction targets. The report noted that effects of 1080 poisoning are temporary; pest populations will eventually rebound.

2. *Can the method increase populations of native species?* The report noted that there is considerable literature to indicate that 1080, when used well, benefits a variety of native species. Birds including kākāriki, kererū, kiwi, kōkako, mōhua, robins, tomtits and whio have responded well to pest control programmes using aerial 1080 operations, with increased chick and adult survival and increased population size. Not all 1080 operations have been successful in populations of native species. Some operations that have failed don't appear to have killed enough pests. Others may have been mis-timed.

3. *Can the method rapidly knock down irrupting populations of pests?* A fast knockdown of pests is often needed in the late winter or early spring to protect birds during nesting season, particularly in years of high seed production from forest trees. The literature suggests different pests respond differently, but aerial 1080 can knock down possum, rat and stoat numbers in areas of any size in 2–3 weeks.

4. *Can the method be used on a large scale in remote areas?* Much of New Zealand's conservation estate consists of vast areas of steep hills and mountains that are difficult to access. Here, the only options are to drop poison from a helicopter, or perhaps use a biocontrol method that would spread itself through predator populations.

5. *Is the method sufficiently cost-effective?* An aerial 1080 operation including pre-feeding can now cost $12–$16 per hectare. Ground control of possums alone (not rats and stoats) in easily accessible farmland can cost as little as $4 per hectare, but as much as $40 per hectare on bush-pasture edge. Costs rise significantly if tracks, bridges and huts are needed for access; in rugged country or areas with difficult vegetation, possum control can cost $80 per hectare or more. In especially rugged terrain, ground control is not possible.

6. *Does the method leave residues in the environment?* In one large analysis, 2537 water samples were analysed for 1080 within 24 hours of an aerial drop. Traces were found in 86 samples, with six having concentrations higher than the Ministry of Health trigger value. Traces of 1080 have been found in soil, though at concentrations 200–500 times lower than that required to kill native insects such as ants and wētā. It takes up to a week for all traces of 1080 to be eliminated from the bodies of poisoned possums.

7. *Can by-kill be minimised?* The report noted that by-kill is almost inevitable with any pest control method. Individuals from 19 species of native birds and 13 species of introduced birds have been found dead after aerial 1080 drops. Most of these recorded bird deaths were associated with four operations 35 years prior, which used poor-quality carrot baits with many small fragments. The method of 1080 application has improved, and by-kill is typically much reduced. Although it is now infrequent, individual aerial 1080 operations still sometimes affect local bird populations if not carried out with sufficient care. One example

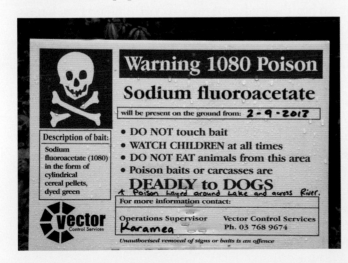

**Warning 1080 Poison**
**Sodium fluoroacetate**
will be present on the ground from: **2 - 9 - 2013**

Description of bait:
Sodium fluoroacetate (1080) in the form of cylindrical cereal pellets, dyed green

- **DO NOT touch bait**
- **WATCH CHILDREN at all times**
- **DO NOT EAT animals from this area**
- Poison baits or carcasses are
**DEADLY to DOGS**
* Poison Layed around Lake and across River.

For more information contact:

| Operations Supervisor | Vector Control Services |
|---|---|
| Karamea | Ph. 03 768 9674 |

*Unauthorised removal of signs or baits is an offence*

**vector** Control Services

A sign warning of 1080 in the Buller District, West Coast of the South Island.
*Roland Knauer / Alamy*

was the deaths of 7 of 17 monitored kea from 1080 poisoning following an aerial operation in South Westland, whereby a helicopter dropped 1080 above the bush-line in kea habitat. Introduced deer and pigs can also be killed after 1080 drops. Wild deer may eat baits, and pigs may eat baits or the carcasses of animals that have eaten baits. The proportion of the deer population that is killed in any operation depends on several factors, including time of year, type of bait used, and whether pre-feeding with non-toxic baits is carried out. Eight dogs are reported to have died from 1080 poisoning over 2007–11. Two died where the operation was not adequately notified, which is a breach of standard operating procedures. Other incidents of dog or other farm animal deaths may not have been reported.

8. *Does the method endanger people?* People could die if they consume enough 1080, either by eating baits or by consuming contaminated food or water. However, in the 60 years of use of 1080 in New Zealand, there are no known records of any deaths from people consuming baits from the field use of 1080. There is one case from New Zealand in the 1960s where it appears a possum hunter died after eating 1080-laced jam bait that was present in his home. A jam matrix is now banned as a form of 1080 application. 1080 residues have never been recorded in public drinking water supplies. The highest recorded concentration in any other water sample following a 1080 operation is 9 parts per billion. At this concentration an adult would need to drink thousands of litres of water at one time to risk death. Repeated and frequent sub-lethal exposure could be harmful for human health. Studies of the effect of 1080 on rats, ferrets, ducks, starlings, lizards and invertebrates have shown that repeated non-lethal doses of 1080 can damage organs such as the heart, muscles and testes. Studies with rats have shown that prolonged exposure to high doses of 1080 may affect the development of unborn young. Similar effects could happen in humans with prolonged exposure to sub-lethal but high doses of 1080.

9. *Does the method kill humanely?* 1080 works by interrupting the body's energy production systems: an animal's cells are starved of energy and subsequently vital functions in the body stop. 1080 acts on different animals in different ways. Herbivores usually die of heart failure, whereas carnivores are more likely to suffer convulsions and respiratory failure. Rats can show pain-related behaviours such as increased grooming and stomach scratching, altered breathing, lack of coordination, and convulsions. Deer have been recorded as becoming lethargic and lying down quietly without convulsions or leg-thrashing. Dogs, stoats and

ferrets have all been observed to go through states of fitting and uncoordinated movement to difficulty in breathing, lethargy and paralysis. Vomiting can also occur. It was noted that humaneness is a subjective measure: different people have different opinions on how humane a particular pest control method is.

The report examined the potential of other alternatives for the control of possums, rats and stoats. Trapping and fur harvesting, other poisons including anticoagulants, and biological control such as contraceptive vaccines were considered. The conclusion was that 1080 was, at the time, the only viable alternative, but research into other controls was strongly encouraged: '1080 is the only poison currently available for aerial pest control on the mainland that can do this job.'*

### Case study 2: Human disease control using genetically modified mosquitoes

The yellow fever mosquito, *Aedes aegypti*, is an invasive species in many tropical regions around the world. In Florida it makes up just 4% of the mosquito population yet is responsible for almost all the mosquito-borne human diseases for the region. Zika, dengue, chikungunya and yellow fever are diseases that can all be transmitted by this mosquito. Pesticides have been heavily used for its control but are becoming infective due to the evolution of resistance in Florida and many countries around the globe.

An increasing human-disease burden saw the Florida Keys Mosquito Control District (FKMCD) approach Oxitec in 2010 to investigate releasing genetically modified mosquitoes in order to suppress or locally eradicate yellow fever mosquitoes. 'Unfortunately, we're seeing our toolbox shrinking due to resistance,' said Andrea Leal of the FKMCD. 'That's one of the reasons why we're really looking at these new innovative tools and new ways to control this mosquito.'[86] The Oxitec programme uses the 'female-specific Release of insects carrying a dominant lethal' (fsRIDL) approach. A genetic modification mean female mosquitoes will die during development, except in the laboratory under special rearing condition. Males survive. But these males, which don't bite, will mate with the wild female mosquitoes that do bite and are responsible for disease transmission. Any female progeny die, but their male offspring become carriers

---

* Full report available at pce.parliament.nz/media/1294/evaluating-the-use-of-1080.pdf

of the gene and pass it to all future generations. As more and more females die, the mosquito populations and human disease prevalence will decline.

The Oxitec programme began in April 2021. If this Florida trial is successful, the US Environmental Protection Agency will consider allowing more releases within the United States. Oxitec developed a substantial public engagement campaign with fliers, public and online talks and events. Public engagement was needed as there was substantial anger and resistance to the releases.

There are and have been concerns about using Oxitec's genetically modified *Aedes aegypti* for disease control. Between 2013 and 2015, 450,000 of these male mosquitoes were released in Brazil each week. Populations of the mosquito consequently fell by 95%.[87] An independent team, however, subsequently raised alarm over the release, publishing a paper in the journal *Scientific Reports*. They state that they found 'clear evidence that portions of the transgenic strain genome have been incorporated into the target population'.[88] The media reports of their study suggested that the release had gone 'horrifically wrong' and had resulted in 'extra-dangerous, mutant blood-suckers flying rampant in a Brazilian city'.[89] An Oxitec spokesperson called the *Scientific Reports* study 'an unqualified research article with misleading, speculative, and unsubstantiated claims and statements about Oxitec's mosquito technology'. The journal editors agreed with Oxitec, issuing a statement noting that the language used in the work was unjustified and that no data was available to support some of their claims.[90] The study had some major flaws. Further, some of the authors hadn't approved the final version of the paper. One report suggested that six of the 10 authors requested the *Scientific Reports* study be retracted.[89]

*Aedes aegypti*, drinking blood. Only females drink blood, to nourish their eggs, but in doing so may transmit disease. *Science Photo Library / Alamy*

Oxitec was contracted by the Florida Keys Mosquito Control District to implement the release of its genetically modified *Aedes aegypti*. *Oxitec*

Nevertheless, the damage to public confidence was largely done. The media reporting of the *Scientific Reports* study probably influenced the Florida public opinion on genetically modified mosquitoes and will do so around the world for years to come.

## Case study 3: Genetically modified corn or maize

The European corn borer (*Ostrinia nubilalis*) in the United States is known as the 'billion-dollar bug'. The moth was unintentionally introduced to and observed in North America from Europe around 1917. Its larvae bore holes into and eat all parts of corn plants and will also eat a range of other crop plants including potatoes and peppers. The insects' feeding allows entry for plant pathogens, and the substantial costs of crop losses and the need for controls by pesticide application has given rise to its pricey nickname. In 1996 genetically modified varieties of corn and maize were released that provided resistance to the European corn borer. The plants have been modified to produce a crystal protein or toxin from a soil bacterium, *Bacillus thuringiensis* (*Bt*). Scientists have known about *Bt* for several decades and used it as a 'natural' or 'biopesticide' that is highly specific to a small range of insects. The proteins or toxin produced will bind only to specific cells and under specific conditions that are found in the midgut of only some insects. Genetically modified varieties of maize with this gene are referred to as *Bt* maize. Genetically modified varieties with resistance to this and other maize pests have been developed since.

Growers in the United States quickly adopted *Bt* maize. They considered it to have 'in the bag' protection against several key insect pests, sometimes describing the plant as 'bullet proof'.[91] In a very short period of time, *Bt* maize adoption increased substantially.[92] *Bt* maize is now grown on 61 million hectares of land around the world, representing nearly a third of the land where genetically modified crops are sown.[79] Insecticide use has declined dramatically, with benefits extending to a range of nearby crops. A meta-analysis of 147 studies concluded:

> On average, GM technology adoption has reduced chemical pesticide use by 37%, increased crop yields by 22%, and increased farmer profits by 68%. Yield gains and pesticide reductions are larger for insect-resistant crops than for herbicide-tolerant crops. Yield and profit gains are higher in developing countries than in developed countries.[93]

The controversies surrounding genetically modified crops focus on human and environmental safety, consumer choice and labelling, food security, intellectual

European corn borer (*Ostrinia nubilalis*) feeding on corn. *Tomasz Klejdysz / Alamy*

property rights, ethics, and poverty reduction. People are concerned about the risks of 'tampering with Mother Nature'.

What specific health concerns do consumers have? Is recombinant technology really beneficial?[94] Despite several reviews concluding that the 'risks associated with GM crops have proven to be low to non-existent',[92,94] people still have anxieties. Human health risks associated with genetically modified foods are associated with toxins, allergens or genetic hazards. Bawa and Anilakumar (2013) write: 'It took us many years to realize that DDT might have oestrogenic activities and affect humans, but we are now being asked to believe that everything is OK with GM foods because we haven't seen any dead bodies.'[94]

In regard to environmental effects with crops such as *Bt* maize, the effects on insect communities appear to result primarily on predators and parasitoids of the European corn borer. A lower number of corn borer results in lower numbers of natural enemies of the moth. Similarly, a lower amount of plant damage has resulted in lower numbers of insects with larvae that feed on decaying organic matter.[91]

A protester holds placards during a protest against genetically modified organisms in the UK in 2013. *Pete Maclaine / Alamy*

## Further reading and discussion

1. Attaran et al. write: 'Malaria kills over one million people, mainly children, in the tropics each year, and DDT remains one of the few affordable, effective tools against the mosquitoes that transmit the disease.' The authors explain that the scientific literature on withdrawing DDT is unpersuasive, and the benefits of DDT in saving lives from malaria are worth the risks. Are they correct? What is the solution to the malaria problem for developing nations?

*See:* Attaran, A. et al. (2000). Balancing risks on the backs of the poor. *Nature Medicine, 6*(7): 729–731. doi.org/10.1038/77438

2. Is 1080 a 'wicked' problem? Perhaps it is 'partially wicked' or even 'post-normal'. How do we discuss 1080 with people who hold strong anti-1080 views? Is a scientific argument the only one we should use?

*See:* Green & Rohan (2012). Opposition to aerial 1080 poisoning for control of invasive mammals in New Zealand: Risk perceptions and agency responses. *Journal of the Royal Society of New Zealand, 42*(3):185–213. doi.org/10.1080/030 36758.2011.556130

3. Are CRISPR-based gene drives a biocontrol silver bullet for conservation or global threat? Should New Zealand use a gene drive for the control of possum or rat pests? What would it take to convince you that we should use this technology? Are you convinced it is a conservation threat?

*See:* Webber et al. (2015). Opinion: Is CRISPR-based gene drive a biocontrol silver bullet or global conservation threat? *Proceedings of the National Academy of Sciences of the USA, 112*(34), 10565–10567. doi.org/10.1073/pnas.1514258112

4. India has seen approximately 300,000 farmer-suicides over the past two decades. Opponents of biotechnology have attributed the increase of suicides to the monopolisation of genetically modified crops, centering on patent control, application of terminator technology, marketing strategy and increased production costs. These heartbreaking deaths are widely reported as a negative outcome of genetically modified crops. What is the evidence? Has it been as bad as it sounds for Indian farmers?

*See:* Thomas & De Tavernier (2017). Farmer-suicide in India: Debating the role of biotechnology. *Life Sciences, Society and Policy, 13*(1), 8. doi.org/10.1186/ s40504-017-0052-z

# References

1. The Parliamentary Commissioner for the Environment. (2011). *Evaluating the use of 1080: Predators, poisons and silent forests.* pce.parliament.nz/media/1294/evaluating-the-use-of-1080.pdf

2. Department of Conservation. (2021, May). *Tiakina Ngā Manu operations.* doc.govt.nz/our-work/tiakina-nga-manu/predator-control-programme/

3. Environmental Risk Management Authority. (2008). *Environmental Risk Management Authority Decision: Application for the reassessment of a hazardous substance under section 63 of the Hazardous Substances and New Organisms Act 1996: Sodium fluoroacetate (1080) and formulated substances containing 1080. Application Number: HRE05002.* epa.govt.nz/assets/FileAPI/hsno-ar/HRE05002/f7659614b3/HRE05002-062.pdf

4. Gudsell, K. (2018, September 18). *DOC staff face more abuse after anti-1080 protests.* Radio New Zealand. rnz.co.nz/news/national/366602/doc-staff-face-more-abuse-after-anti-1080-protests

5. Brown, K., Elliott, G., Innes, J., et al. (2015). *Ship rat, stoat and possum control on mainland New Zealand: An overview of techniques, successes and challenges.* Department of Conservation. doc.govt.nz/Documents/conservation/threats-and-impacts/animal-pests/ship-rat-stoat-possum-control.pdf

6. Stern, V. M., Smith, R. F., van den Bosch, R., et al. (1959). The integration of chemical and biological control of the spotted alfalfa aphid: The integrated control concept. *Hilgardia, 29*(2), 81–101. doi.org/10.3733/hilg.v29n02p081

7. Pedigo, L. P., Hutchins, S. H., & Higley, L. G. (1986). Economic injury levels in theory and practice. *Annual Review of Entomology, 31*, 341–368. doi.org/10.1146/annurev.en.31.010186.002013

8. The Food and Agriculture Organization of the United Nations and World Health Organization. (2013). *International code of conduct on the distribution and use of pesticides: Guidelines on data requirements for the registration of pesticides.* apps.who.int/iris/bitstream/handle/10665/337301/WHO-HTM-NTD-WHOPES-2013.7-eng.pdf

9. Attaran, A., Roberts, D. R., Curtis, C. F., et al. (2000). Balancing risks on the backs of the poor. *Nature Medicine, 6*, 729–731. doi.org/10.1038/77438

10. Rogan, W. J., & Chen, A. (2005). Health risks and benefits of bis(4-chlorophenyl)-1,1,1-trichloroethane (DDT). *The Lancet, 366*(9487), 763–773. doi.org/10.1016/s0140-6736(05)67182-6

11. Carson, R. (1962). *Silent spring.* Houghton Mifflin Company.

12. Stokstad, E. (2007). Species conservation: Can the bald eagle still soar after it is delisted? *Science, 316*(5832), 1689–1690. doi.org/10.1126/science.316.5832.1689

13. McLaughlin, D. (2010). *Fooling with nature: Silent Spring revisited.* PBS Frontline. pbs.

org/wgbh/pages/frontline/shows/nature/disrupt/sspring.html

14. Gay, H. (2012). Before and after *Silent Spring*: From chemical pesticides to biological control and integrated pest management – Britain, 1945–1980. *Ambix, 59*(2), 88–108. doi.org/10.1179/174582312X13345259995930

15. Geisz, H. N., Dickhut, R. M., Cochran, M. A., et al. (2008). Melting glaciers: A probable source of DDT to the Antarctic marine ecosystem. *Environmental Science and Technology, 42*(11), 3958–3962. doi.org/10.1021/es702919n

16. Cirillo, P. M., La Merrill, M. A., Krigbaum, N. Y., et al. (2021). Grandmaternal perinatal serum DDT in relation to granddaughter early menarche and adult obesity: Three generations in the child health and development studies cohort. *Cancer Epidemiology Biomarkers & Prevention, 2021*(30), 1480–1488. doi.org/10.1158/1055-9965.Epi-20-1456

17. van den Berg, H., Zaim, M., Yadav, R. S., et al. (2012). Global trends in the use of insecticides to control vector-borne diseases. *Environmental Health Perspectives, 120*(4), 577–582. doi.org/10.1289/ehp.1104340

18. van den Berg, H., Manuweera, G., & Konradsen, F. (2017). Global trends in the production and use of DDT for control of malaria and other vector-borne diseases. *Malaria Journal, 16*, 401. doi.org/10.1186/s12936-017-2050-2

19. Atwood, D., & Paisley-Jones, C. (2017). *Pesticides industry sales and usage: 2008–2012 market estimates*. United States Environmental Protection Agency. epa.gov/sites/production/files/2017-01/documents/pesticides-industry-sales-usage-2016_0.pdf

20. Van Bruggen, A. H. C., He, M. M., Shin, K., et al. (2018). Environmental and health effects of the herbicide glyphosate. *Science of the Total Environment, 616–617*, 255–268. doi.org/10.1016/j.scitotenv.2017.10.309

21. Catarzi, D., Colotta, V., & Varano, F. (2007). Competitive AMPA receptor antagonists. *Medicinal Research Reviews, 27*(2), 239–278. doi.org/10.1002/med.20084

22. Owen, M. D. K., Beckie, H. J., Leeson, J. Y., et al. (2015). Integrated pest management and weed management in the United States and Canada. *Pest Management Science, 71*(3), 357–376. doi.org/10.1002/ps.3928

23. Rennie, R. (2021, August 12). Herbicide resistance surging. *Farmers Weekly.* farmersweekly.co.nz/section/agribusiness/view/herbicide-resistance-surging

24. McComb, B. C., Curtis, L., Chambers, C. L., et al. (2008). Acute toxic hazard evaluations of glyphosate herbicide on terrestrial vertebrates of the Oregon coast range. *Environmental Science and Pollution Research, 15*(3), 266–272. doi.org/10.1065/espr2007.07.437

25. Motta, E. V. S., Raymann, K., & Moran, N. A. (2018). Glyphosate perturbs the gut microbiota of honey bees. *Proceedings of the National Academy of Sciences, 115*(41), 10305–10310. doi.org/10.1073/pnas.1803880115

26. Hawes, C., Haughton, A. J., Osborne, J. L., et al. (2003). Responses of plants and invertebrate trophic groups to contrasting herbicide regimes in the Farm Scale

Evaluations of genetically modified herbicide-tolerant crops. *Philosophical Transactions of the Royal Society B: Biological Sciences, 358*(1439), 1899–1913. doi.org/10.1098/rstb.2003.1406

27. Douwes, J., 't Mannetje, A., McLean, D., et al. (2018). Carcinogenicity of glyphosate: Why is New Zealand's EPA lost in the weeds? *New Zealand Medical Journal, 131*(1472), 82–89.

28. Illien, N. (2021, June 13). Swiss voters reject proposal to ban synthetic pesticides. *New York Times*. nytimes.com/2021/06/13/world/swiss-pesticide-referendum-ban.html

29. Food and Agriculture Organization of the United Nations. (2021, June 27). *Pesticides use database*. fao.org/faostat/en/#data/RP

30. Schulz, R., Bub, S., Petschick, L. L., et al. (2021). Applied pesticide toxicity shifts toward plants and invertebrates, even in GM crops. *Science, 372*(6537), 81–84. doi.org/10.1126/science.abe1148

31. Kranthi, K. R., & Stone, G. D. (2020). Long-term impacts of *Bt* cotton in India. *Nature Plants, 6*(3), 188–196. doi.org/10.1038/s41477-020-0615-5

32. Chandler, D., Bailey, A. S., Tatchell, G. M., et al. (2011). The development, regulation and use of biopesticides for integrated pest management. *Philosophical Transactions of the Royal Society B: Biological Sciences, 366*(1573), 1987–1998. doi.org/10.1098/rstb.2010.0390

33. Phillips, C. B., Brown, K., Green, C., et al. (2020). Eradicating the large white butterfly from New Zealand eliminates a threat to endemic Brassicaceae. *PLoS One, 15*(8), Article e0236791. doi.org/10.1371/journal.pone.0236791

34. Morandin, L. A., Winston, M. L., Franklin, M. T., et al. (2005). Lethal and sub-lethal effects of spinosad on bumble bees (*Bombus impatiens* Cresson). *Pest Management Science, 61*(7), 619–626. doi.org/10.1002/ps.1058

35. Mitter, N., Worrall, E. A., Robinson, K. E., et al. (2017). Clay nanosheets for topical delivery of RNAi for sustained protection against plant viruses. *Nature Plants, 3*, Article 16207. doi.org/10.1038/nplants.2016.207

36. San Miguel, K., & Scott, J. G. (2016). The next generation of insecticides: dsRNA is stable as a foliar-applied insecticide. *Pest Management Science, 72*(4), 801–809. doi.org/10.1002/ps.4056

37. Heath, G., Childs, D., Docker, M. F., et al. (2014). RNA interference technology to control pest sea lampreys – A proof-of-concept. *PLoS One, 9*(2), Article e88387. doi.org/10.1371/journal.pone.0088387

38. Horak, K. E. (2020). RNAi: Applications in vertebrate pest management. *Trends in Biotechnology, 38*(11), 1200–1202. doi.org/10.1016/j.tibtech.2020.05.001

39. Clapperton, B. K., Murphy, E. C., & Razzaq, H. A. A. (2017). *Mammalian pheromones – new opportunities for improved predator control in New Zealand* (Science for Conservation, issue 300). Department of Conservation. doc.govt.nz/globalassets/

documents/science-and-technical/sfc330entire.pdf

40. El-Sayed, A. M., Suckling, D. M., Wearing, C. H., et al. (2006). Potential of mass trapping for long-term pest management and eradication of invasive species. *Journal of Economic Entomology, 99*(5), 1550–1564. doi.org/10.1093/jee/99.5.1550

41. Bray, D. P., Alves, G. B., Dorval, M. E., et al. (2010). Synthetic sex pheromone attracts the leishmaniasis vector *Lutzomyia longipalpis* to experimental chicken sheds treated with insecticide. *Parasites and Vectors, 3*, Article 16. doi.org/10.1186/1756-3305-3-16

42. Brockerhoff, E. G., Suckling, D. M., Kimberley, M., et al. (2012). Aerial application of pheromones for mating disruption of an invasive moth as a potential eradication tool. *PLoS One, 7*(8), Article e43767. doi.org/10.1371/journal.pone.0043767

43. Russell, J. C., Towns, D. R., Anderson, S. H., et al. (2005). Intercepting the first rat ashore. *Nature, 437*(7062), 1107. doi.org/10.1038/4371107a

44. Robinson, S. A., & Copson, G. R. (2014). Eradication of cats (*Felis catus*) from subantarctic Macquarie Island. *Ecological Management & Restoration, 15*(1), 34–40. doi.org/10.1111/emr.12073

45. Keane, R., & Crawley, M. J. (2002). Exotic plant invasions and the enemy release hypothesis. *Trends in Ecology & Evolution, 17*(4), 164–170. doi.org/10.1016/s0169-5347(02)02499-0

46. Hokkanen, H. M. T., & Pimentel, D. (1989). New associations in biological control: Theory and practice. *The Canadian Entomologist, 121*(10), 829–840. doi.org/10.4039/Ent121829-10

47. Roy, H. E., Brown, P. M. J., Adriaens, T., et al. (2016). The harlequin ladybird, *Harmonia axyridis*: Global perspectives on invasion history and ecology. *Biological Invasions, 18*(4), 997–1044. doi.org/10.1007/s10530-016-1077-6

48. Vilcinskas, A., Stoecker, K., Schmidtberg, H., et al. (2013). Invasive harlequin ladybird carries biological weapons against native competitors. *Science, 340*(6134), 862–863. doi.org/10.1126/science.1234032

49. Simberloff, D., & Stiling, P. (1996). Risks of species introduced for biological control. *Biological Conservation, 78*(1–2), 185–192. doi.org/10.1016/0006-3207(96)00027-4

50. van Lenteren, J. C., Bolckmans, K., Köhl, J., et al. (2018). Biological control using invertebrates and microorganisms: Plenty of new opportunities. *BioControl, 63*, 39–59. doi.org/10.1007/s10526-017-9801-4

51. Morin, L. (2020). Progress in biological control of weeds with plant pathogens. *Annual Review of Phytopathology, 58*, 201–223. doi.org/10.1146/annurev-phyto-010820-012823

52. Gurr, G. M., Wratten, S. D., Landis, D. A., et al. (2017). Habitat management to suppress pest populations: Progress and prospects. *Annual Review of Entomology, 62*, 91–109. doi.org/10.1146/annurev-ento-031616-035050

53. Gurr, G. M., Lu, Z., Zheng, X., et al. (2016). Multi-country evidence that crop

diversification promotes ecological intensification of agriculture. *Nature Plants, 2,* Article 16014. doi.org/10.1038/nplants.2016.14

54. Terraube, J. (2019). Can protected areas mitigate Lyme disease risk in Fennoscandia? *Ecohealth, 16*(2), 184–190. doi.org/10.1007/s10393-019-01408-4

55. Zimmermann, H. G., Moran, V. C., & Hoffmann, J. H. (2000). The renowned cactus moth, *Cactoblastis cactorum*: Its natural history and threat to native *Opuntia* floras in Mexico and the United States of America. *Diversity and Distributions, 6*(5), 259–269. doi.org/10.1046/j.1472-4642.2000.00088.x

56. Biere, A., & Bennett, A. E. (2013). Three-way interactions between plants, microbes and insects. *Functional Ecology, 27*(3), 567–573. doi.org/10.1111/1365-2435.12100

57. Barratt, B. I. P., Howarth, F. G., Withers, T. M., et al. (2010). Progress in risk assessment for classical biological control. *Biological Control, 52*(3), 245–254. doi.org/10.1016/j.biocontrol.2009.02.012

58. Bale, J. S., van Lenteren, J. C., & Bigler, F. (2008). Biological control and sustainable food production. *Philosophical Transactions of the Royal Society B: Biological Sciences, 363*(1492), 761–776. doi.org/10.1098/rstb.2007.2182

59. Paynter, Q., Konuma, A., Dodd, S. L., et al. (2017). Prospects for biological control of *Lonicera japonica* (Caprifoliaceae) in New Zealand. *Biological Control, 105*, 56–65. doi.org/10.1016/j.biocontrol.2016.11.006

60. Hazardous Substances and New Organisms Act 1996. legislation.govt.nz/act/public/1996/0030/latest/DLM381222.html

61. Gullan, P. J., & Cranston, P. S. (2010). *The insects: An outline of entomology* (4th edn.). Wiley-Blackwell.

62. Walker, J. T. S., Suckling, D. M., & Wearing, C. H. (2017). Past, present, and future of integrated control of apple pests: The New Zealand experience. *Annual Review of Entomology, 62*, 231–248. doi.org/10.1146/annurev-ento-031616-035626

63. St. George, D. A., & Johnson, M. D. (2021). Effects of habitat on prey delivery rate and prey species composition of breeding barn owls in winegrape vineyards. *Agriculture, Ecosystems & Environment, 312.* doi.org/10.1016/j.agee.2021.107322

64. Mahlaba, T. A. M., Monadjem, A., McCleery, R., et al. (2017). Domestic cats and dogs create a landscape of fear for pest rodents around rural homesteads. *PLoS One, 12*(2), Article e0171593. doi.org/10.1371/journal.pone.0171593

65. Leftwich, P. T., Bolton, M., & Chapman, T. (2016). Evolutionary biology and genetic techniques for insect control. *Evolutionary Applications, 9*(1), 212–230. doi.org/10.1111/eva.12280

66. Hendrichs, J., & Robinson, A. (2009). Sterile insect technique. In V. H. Resh, & R. T. Carde (eds.). *Encyclopedia of insects* (2nd edn., pp. 953–957). Academic Press, Elsevier Science Publisher. doi.org/10.1016/B978-0-12-374144-8.00252-6

67. Gemmell, N. J., Jalilzadeh, A., Didham, R. K., et al. (2013). The Trojan female technique: A novel, effective and humane approach for pest population control. *Proceedings of the Royal Society B: Biological Sciences, 280*(1773), Article 20132549. doi. org/10.1098/rspb.2013.2549

68. Zheng, X., Zhang, D., Li, Y., et al. (2019). Incompatible and sterile insect techniques combined eliminate mosquitoes. *Nature, 572*(7767), 56–61. doi.org/10.1038/s41586-019-1407-9

69. Walker, T., Johnson, P. H., Moreira, L. A., et al. (2011). The wMel *Wolbachia* strain blocks dengue and invades caged *Aedes aegypti* populations. *Nature, 476*(7361), 450–453. doi.org/10.1038/nature10355

70. Utarini, A., Indriani, C., Ahmad, R. A., et al. (2021). Efficacy of Wolbachia-infected mosquito deployments for the control of dengue. *The New England Journal of Medicine, 384*(23), 2177–2186. doi.org/10.1056/NEJMoa2030243

71. Gever, J. (2020, November 20). *Modified mosquitoes suppress dengue in field trial.* MedPage Today. medpagetoday.com/meetingcoverage/astmh/89817

72. New biocontrol research to help prevent mice plagues. (2021, June 4). *The National Tribune, Australia.* nationaltribune.com.au/new-biocontrol-research-to-help-prevent-mice-plagues/

73. PETA Australia. (2021, June 5). Update: NSW Government invests in mice control. peta.org.au/news/mice-bio-control/

74. Webber, B. L., Raghu, S., & Edwards, O. R. (2015). Opinion: Is CRISPR-based gene drive a biocontrol silver bullet or global conservation threat? *Proceedings of the National Academy of Sciences, 112*(34), 10565–10567. doi.org/10.1073/pnas.1514258112

75. Lester, P. J., Bulgarella, M., Baty, J. W., et al. (2020). The potential for a CRISPR gene drive to eradicate or suppress globally invasive social wasps. *Scientific Reports, 10*(1), Article 12398. doi.org/10.1038/s41598-020-69259-6

76. Lester, P. J. (2018). *The vulgar wasp: The story of a ruthless invader and ingenious predator.* Victoria University Press.

77. Gonsalves, D. (1998). Control of papaya ringspot virus in papaya: A case study. *Annual Review of Phytopathology, 36*, 415–437. doi.org/10.1146/annurev.phyto.36.1.415

78. Voosen, P. (2011, September 21). Crop savior blazes biotech trail, but few scientists or companies are willing to follow. *New York Times.* archive.nytimes.com/www. nytimes.com/gwire/2011/09/21/21greenwire-crop-savior-blazes-biotech-trail-but-few-scien-88379.html

79. International Service for the Acquisition of Agri-biotech Applications. (2019). *Global status of commercialized biotech/gm crops: 2019 (ISAAA Briefs No. 55, International Service for the Acquisition of Agri-Biotech Applications, Ithaca, NY, 2019).* isaaa.org/resources/publications/briefs/55/executivesummary/default.asp

80. Wu, K.-M., Lu, Y.-H., Feng, H.-Q., et al. (2008). Suppression of cotton bollworm in

multiple crops in China in areas with *Bt* toxin-containing cotton. *Science, 321*(5896), 1676–1678. doi.org/10.1126/science.1160550

81. Hudson, M., Mead, A. T. P., Chagné, D., et al. (2019). Indigenous perspectives and gene editing in Aotearoa New Zealand. *Frontiers in Bioengineering and Biotechnology, 7*, 70. doi.org/10.3389/fbioe.2019.00070

82. Saturday Morning. (2021, June 12). Tame Malcolm: Using indigenous methods to fight pests. RNZ. rnz.co.nz/national/programmes/saturday/audio/2018799442/tame-malcolm-using-indigenous-methods-to-fight-pests

83. Cressey, D. (2015). Widely used herbicide linked to cancer. *Nature*. doi.org/10.1038/nature.2015.17181

84. Lester, P. J. (2020). *Healthy bee, sick bee: The influence of parasites, pathogens, predators and pesticides on honey bees*. Victoria University Press.

85. Byrom, A. E., Innes, J. & Binny, R. N. (2016). A review of biodiversity outcomes from possum-focused pest control in New Zealand. *Wildlife Research, 43*, 228–253. doi.org/10.1071/WR15132

86. Waltz, E. (2021). First genetically modified mosquitoes released in the United States. *Nature, 593*(7858), 175–176. doi.org/10.1038/d41586-021-01186-6

87. Carvalho, D. O. et al. (2015). Suppression of a field population of *Aedes aegypti* in Brazil by sustained release of transgenic male mosquitoes. *PLOS Neglected Tropical Diseases, 9*(7), e0003864. doi.org/10.1371/journal.pntd.0003864

88. Evans, B. R. et al. (2019). Transgenic *Aedes aegypti* mosquitoes transfer genes into a natural population. *Scientific Reports, 9*, Article 13047. doi.org/10.1038/s41598-019-49660-6

89. Mole, B. (2019, October 3). Study claimed a GMO trial went horrifically wrong. The study's authors disagree. *ARS Technica*. arstechnica.com/science/2019/10/study-claimed-a-gmo-trial-went-horrifically-wrong-the-studys-authors-disagree/

90. Evans, B. R. et al. (2020). Editorial expression of concern: Transgenic *Aedes aegypti* mosquitoes transfer genes into a natural population. *Scientific Reports, 10*, Article 5524. nature.com/articles/s41598-020-62398-w

91. Hellmich, R. L. & Hellmich, K. A. (2012). Use and impact of *Bt* maize. *Nature Education Knowledge, 3*(10), 4. dr.lib.iastate.edu/handle/20.500.12876/23764

92. Carzoli, A. K., Aboobucker, S. I., Sandall, L. L., Lübberstedt, T. T. & Suza, W. P. (2018). Risks and opportunities of GM crops: *Bt* maize example. *Global Food Security 19*, 84–91. doi.org/10.1016/j.gfs.2018.10.004.

93. Klumper, W. & Qaim, M. (2014). A meta-analysis of the impacts of genetically modified crops. *PLoS One, 9*, e111629. doi.org/10.1371/journal.pone.0111629

94. Bawa, A. S. & Anilakumar, K. R. (2013). Genetically modified foods: Safety, risks and public concerns – a review. *Journal of Food Science and Technology, 50*, 1035–1046. doi.org/10.1007/s13197-012-0899-1

# 6. PANDEMICS

## The source and evolution of pandemic pathogens, their spread and control

We have entered an era of pandemics. The frequency of pandemics is rising, largely driven by an increasing incidence of emerging disease events. We now see more than five new diseases emerging in humans every year, each with the potential to become a pandemic.[1] Nearly all of these diseases, including COVID-19, Ebola and HIV/AIDS, are driven by the spillover of pathogens from animals after contact between wildlife, livestock and people. Pathogens have helped shape our history and will play a major role in our future.

An epidemic is the rapid spread of disease to a large number of people in a given population within a short period of time. A pandemic is broadly defined as an epidemic of infectious disease that has spread across large geographic areas, or many regions or countries, and affected substantial numbers of people. Let's look at two pathogens responsible for some of the most catastrophic pandemics in our history (Fig. 6.1).

The Black Death may have killed as many as 200 million people – one-third of the entire European population – over 1347–53. The pathogen responsible was the plague bacterium *Yersinia pestis*.

Doctor Schnabel ('Dr Beak') is a well-known representation of the bubonic plague. Plague doctors practised bloodletting by putting frogs or leeches on the buboes to 'rebalance the humors'. The mask was to keep away bad smells, known as miasma, which were thought to be the principal cause of the disease. The beak was often stuffed with aromatic items like dried flowers or herbs. *Trustees of the British Museum / Wikimedia Commons*

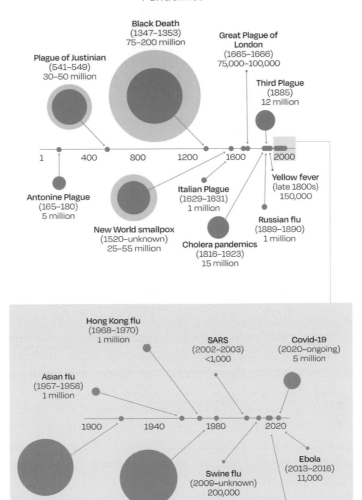

Fig. 6.1: Major pandemics from 1 BCE to the present day. The size of the circle is relative to the number of deaths attributed to each pandemic. The records of early pandemics are uncertain (represented by the lighter shaded circle). The Black Death is the pandemic with the highest known mortality, killing one third of the European population. It is thought to have entered Europe from Crimea in fleas on the black rats that travelled on slave ships. The New World smallpox epidemic had fewer overall deaths, but was even more devastating: the introduction of *Variola* major and *Variola* minor into South America is estimated to have caused the deaths of 90–95% of Indigenous peoples.

A wide variety of rodents serve as the reservoir host species for the disease, which is vectored by fleas. Scientists now believe they have traced the evolution of this bacteria.[2] It is thought to have evolved from a related bacteria as recently as 5700–6000 years ago. An analysis of DNA from plague bacteria on the teeth of nine 3800-year-old human skeletons in Russia showed that the pathogen had quickly lost several genes, but also gained genetic adaptations that enabled it to be hosted and transmitted by fleas.[3] Approximately 1500 years ago, the pathogen acquired additional genes that enabled it to enter lung tissue, to be spread by coughs and sneezes. These newly acquired genes further mutated. A slight change of just one amino acid meant the bacteria could spread deep into tissue after the bite of a flea or rodent.[4] This new and improved *Yersinia pestis* genotype is known from the Justinian Plague of 541–49, one of the world's first major pandemics.

Scholars believe that after the 1347–53 pandemic, 'society became more violent precisely because of the plague, that the mass mortality cheapened life and thus increased warfare, crime, popular revolt, waves of flagellants, and persecutions against the Jews'.[5] The plague has since been shipped around the world with rats and other rodents. In biological warfare during World War II, Japanese scientists spread the plague over Chinese cities. Soldiers performed cruel experiments in which they infected and dissected live prisoners without anaesthetic.[6] And the plague is still with us today. Since 2013 we've seen plague outbreaks in Madagascar and the Democratic Republic of Congo. Thirty flea species are now known to vector the plague to over 200 different mammal hosts.[7] It can be epizootic (epidemic in animal populations) and decimate wildlife. One plague epizootic in the black-tailed prairie dog (*Cynomys ludovicianus*) that began in 2003 reduced a population by 95%.[8]

World War I ended in 1918, having claimed 18 million lives. Nearly three times that number would then be claimed by a pandemic of an influenza A virus that followed. At least 50 million people died as the virus swept the globe. This influenza pandemic has been referred to as 'the greatest medical holocaust in history', as it 'ranks with the plague of Justinian and the Black Death as one of the three most destructive human epidemics'.[9]

The natural reservoir of influenza A viruses is birds, mainly shorebirds and waterfowl. Infections in humans, mammals and poultry result from host-switching events.[10]

New Zealand recorded approximately 8600 deaths from this virus. Māori communities experienced a death rate of 42.3 per thousand people, which is a rate of mortality more than seven times higher than for Europeans. It has been hypothesised that many Māori were immunologically naïve to flu and colds, but also that such high mortality rates were due to Māori living with lower standards of housing, clothing and nourishment.[11] There were two other striking aspects of the mortality data from New Zealand.

First, mortality from influenza A viruses, and many pathogens, is typically highest for the very old and very young. Unusually, in 1918 we saw a high rate of mortality among young adults aged 30–34 (Fig. 6.2). This is a trend also seen overseas, which some hypothesise might be due to older cohorts having a degree of immunity from exposure to a previous virus. (See 'Case study 1: Finding and reconstructing the H1N1 "Spanish flu" pandemic virus', page 244.) The second unusual aspect of the New Zealand data was that the rates of mortality were substantially higher for young adult males than for females. This pattern of mortality is unexplained and unique to New Zealand.

Some communities fared better than others. On the East Cape, locals armed with shotguns manned roadblocks that ensured an effective quarantine. Face

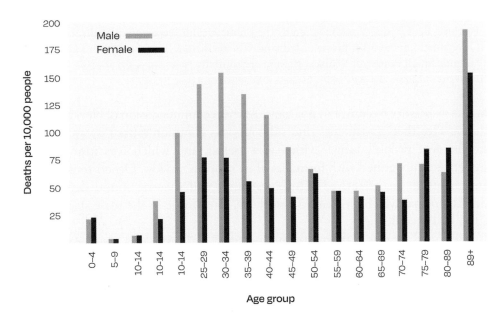

Fig. 6.2: The age-specific death rates for the Pākehā population from the 1918 influenza A virus pandemic. Data for Māori and other ethnic groups is not considered reliable. *Rice (2005)[11]*

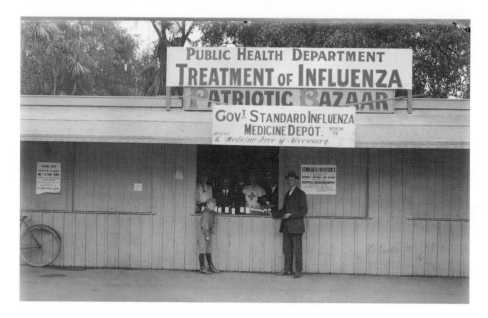

Central Medicine Depot, Cathedral Square, Christchurch, December 1918. The former 'Patriotic Bazaar' that was converted by the Public Health Department to dispense 'Government Standard Influenza Medicine'. The medicines listed were 'Stimulants for Patients. Small Bottles of Whisky, Brandy or Stout'. *The Press (Newspaper): Negatives. Ref: 1/1-008542-G. Alexander Turnbull Library, Wellington, New Zealand. /records/22889273*

masks were worn by some but not most. At the recommendation of New Zealand's Public Health Department, zinc sulphate was commonly used as an inhalant to disinfect people's lungs, which was useless, and which was unfortunately sometimes substituted with formalin when supplies ran low. Others took quinine and swore it was effective.[11] These responses and rhetoric sound depressingly similar to that which we heard from the White House in the early stages of the COVD-19 pandemic.

How can we stop our past from repeating? As I write this chapter we are in the grasp of yet another pandemic. The SARS-CoV-2 virus and the COVID-19 disease it causes have killed more than 6 million people. Can we identify future pathogens? How do they enter human populations, evolve and spread? And what can we do to stem their flow and limit their impact? These are especially relevant questions in a pandemic era.

## The pathogens that cause disease and pandemics

Humans represent a mobile ecosystem – we will never walk alone. A typical person has a total of about 30 trillion cells but carries approximately 38 trillion bacteria. So numerically, you are more bacteria than you are human. But because of their small size, your bacteria contribute only about 0.2 kilograms to your total body mass.[12] You can add viruses, fungi and many other organisms to this 38 trillion. Some of these microbes appear essential to your health and well-being. Others, however, are pathogens. Pathogens are defined as organisms that cause disease to their hosts, with the severity of the disease symptoms referred to as virulence.[13] They are taxonomically diverse, comprising bacteria, viruses, fungi and an array of other unicellular and multicellular organisms.

Approximately 1399 species have been described as human pathogens (Table 6.1). Some are facultative pathogens: they only occasionally cause disease, often when a host's immune system is weakened or compromised, such as the bacteria *Escherichia coli*. Others are obligatory pathogens, as they require damaging a host to fulfil their lifecycle. The illnesses that these obligatory pathogens cause may

| Pathogen type | Number of species | New since 1980 |
|---|---|---|
| Total | 1399 | 87 |
| Taxonomic group | | |
| Bacteria | 541 (39%) | 11 (13%) |
| Fungi | 325 (23%) | 13 (15%) |
| Helminths | 285 (20%) | 1 (1%) |
| Prions | 2 (0.1%) | 1 (1%) |
| Protozoa | 57 (4%) | 3 (3%) |
| Viruses | | |
| DNA viruses | 36 (3%) | 9 (10%) |
| RNA viruses | 153 (11%) | 49 (56%) |

Table 6.1: Numbers of pathogen species that infect humans, by taxonomic category. All records are shown in the first column, with those that have only recently been discovered in the second. These 'new' species can be viewed as emerging infectious diseases (EIDs). Bacterial pathogens were historically prevalent, but RNA viruses (e.g. SARS-CoV-2) represent the majority of new species.
*Data from Woolhouse & Gaunt (2007).*[14]

be due to the direct damage of cells or tissues during bacterial replication, often through the production of toxins that are among the deadliest known poisons, including anthrax. Obligatory pathogens often also induce excessive immune responses that indiscriminately kill infected and uninfected cells. This damage to host tissues can be extensive enough to kill. In the 1918–19 influenza pandemic, for example, the young and otherwise healthy had high rates of mortality, which has been hypothesised to be due to an excessive and strong immune response. They probably died from 'cytokine storms' in their lungs that resulted in sufferers literally drowning in their own body fluids.[13]

Humans have had a long relationship with pathogens. One insight has been gained through a close examination of fossilised human poop. Some of the oldest fossilised human faeces, which were deposited nearly 30,000 years ago, have been found in France and have been found to contain eggs of the nematode roundworm *Ascaris lumbricoides*. After our ancestors began farming livestock, they increased their contact with and chance of infection from domesticated animals. It seems likely that smallpox entered the human population from camels, and mumps from pigs. Our human settlements provided an ideal habitat for other species that serve as reservoir hosts for pathogens, including mice and rats, and vector species, fleas and mosquitoes. Our interactions with both domesticated animals and wildlife have resulted in zoonotic infectious diseases or zoonoses: diseases that are caused by pathogens transmissible from animals to humans. Bacteria represent the largest fraction of pathogens known to infect humans, followed by fungal species, parasitic helminth worms, viruses and protozoa. About 60% of all these pathogens are zoonotic and are spread through human contact with animals. However, this estimate doesn't include the ancient diseases that may have come from animals, such as smallpox and measles, or even the recent ones such as HIV/AIDS that we know came from chimpanzees, because they are no longer transmitted between humans and animals.[15]

The last few decades have seen a change in patterns of pathogen origin and zoonotic spillover from animal hosts. Historically, bacterial pathogens were of prime importance in our list of 1399 human pathogens. These bacterial pathogens include tuberculosis (*Mycobacterium tuberculosis*), which is still prevalent and is a leading cause of human mortality today. *Yersinia pestis* caused the Black Death and is still common in sub-Saharan Africa and Madagascar. Another is cholera, caused by the bacterium *Vibrio cholerae*. Six cholera pandemics, which claimed at least 15 million lives, occurred between 1817 and 1923. Cholera also continues as a form of pestilence with frequent outbreaks, including an ongoing epidemic in Yemen that stems from war and poor sanitation.

Cholera has a special place in our understanding of pandemics and their control. The importance of hygiene and sanitation was first demonstrated by the British physician John Snow, when he disabled a pump supplying infected water during an 1854 cholera outbreak in London by simply removing the handle. Oral or injectable vaccines are another key tool. The microbiologist Louis Pasteur experimented with vaccines for cholera in 1877. Interestingly, most cholera epidemics appear self-limiting, meaning they have often ended without human intervention. Phages appear to be key to this limitation. Phages are viruses that attack bacteria, with Vibriophages specialising in cholera bacteria. Patterns of phage abundance and predation of cholera bacteria may explain the self-limiting nature of seasonal cholera epidemics in Bangladesh and other countries.[16]

While populations around the globe still suffer from these familiar bacterial pathogens, over the last several decades we have been experiencing a surge in infectious disease morbidity and mortality. This surge suggests the onset of a new epidemiologic transition with pathogens termed 'emerging infectious diseases' (EIDs). These 'new' diseases are defined as 'those that have recently increased in incidence, impact, or geographic or host range; that are caused by pathogens that have recently evolved; that are newly discovered; or that have recently changed their clinical presentation'.[17] The world is suffering from one of these currently, with the SARS-CoV-2 virus that causes COVID-19. We didn't know this virus existed before 2019.

An Angolan free-tailed bat. *Ivan Kuzmin / Alamy*

In 2020, the WHO listed EIDs of epidemic concern. They included COVID-19, Crimean-Congo haemorrhagic fever, Ebola and Marburg viruses, Lassa fever, Middle East respiratory syndrome (MERS) and severe acute respiratory syndrome (SARS) coronaviruses, Nipah virus and henipaviral diseases, Rift Valley fever, Zika and 'Disease X' ('a serious international epidemic could be caused by a pathogen currently unknown to cause human disease').[17] This list highlights how RNA viruses have become an increasing problem. RNA viruses accounted for nearly half the EIDs that have emerged since 1980 (Table 6.1), but *all* pathogens on the WHO list of EIDs of epidemic concern are RNA viruses. The exception is, of course, Disease X, but I'd bet a lot of money that it will be an RNA virus (see 'Case study 2: The 2020 IPBES report on biodiversity and pandemics', page 247).

The WHO list is also interesting because seven of the eleven pathogens have their origins in bats.

## How these pathogens arise and evolve

Ebola virus disease is caused by three RNA viruses in the genus Ebolavirus, which were first discovered in 1976. The most dangerous is considered to be the *Zaire ebolavirus* (the Ebola virus). The world's worst Ebola outbreak in Western Africa in 2013–16 was the most widespread, resulting in at least 11,000 deaths spread across Guinea, Liberia, Sierra Leone, Nigeria and Mali. The true fatality toll from this pandemic is likely to be much higher, though we are unable to estimate just how much so. The case fatality rate for Ebola is typically estimated by scientists at 50–63%,[18] but some estimates go as high as 88%,[19] meaning that if you get this disease you are more likely to die than to survive.

Scientists believe they know the specific village, tree and child where this Ebola outbreak began. Emile Ouamouno was a young boy, perhaps just two years old, who died in December 2013 in the village of Méliandou, Guinea. Children would play in a cola tree located approximately 50 metres from Emile's home. It was a roost for a colony of the insectivorous Angolan free-tailed bats (*Mops condylurus*). The bats may have been attracted to the tree and region because of agricultural activity and were a frequent target for children, who would hunt and grill them over fires. A week after Emile's death, the Ebola virus killed his mother, his sister and then his grandmother.

The hollow cola tree that housed a colony of insectivorous Angolan free-tailed bats, considered the likely reservoir source of the Ebola virus. *Saéz et al. (2014)[19]*

The initial symptoms of Ebola virus disease include fever, headache, muscle pain and chills. Later, increasingly ill individuals experience internal or external bleeding, and vomiting or coughing blood. Death is typically caused by multi-organ failure and shock. Viral transmission is through contact with infected fluids, with even the recently deceased a source of viral transmission. A proportion of people who catch this disease recover, but unfortunately their recovery represents a problem because survivors can still harbour the virus. It hides in places where the immune system cannot reach, including the eyes and testes. A small resurgence of the Ebola virus in Guinea in 2016 was attributed to a man who had recovered from an infection more than 500 days prior, who had passed the disease to his wife during sexual intercourse. As Kupferschmidt (2021) has written, the 'Ebola virus may lurk in survivors for many years.'[20]

This scenario for how the Ebola disease arose and spread is common to many diseases and pandemics. There is a substantial pool of pathogens circulating in the world's biodiversity. For example, one study estimated that there are a minimum of 320,000 viruses that infect mammals.[21] The recent 'Workshop Report on Biodiversity and Pandemics of the Intergovernmental Platform on Biodiversity and Ecosystem Services' (IPBES) estimated that an even higher 1.7

million undiscovered viruses exist in mammalian and avian hosts. (See 'Case study 2: The 2020 IPBES report on biodiversity and pandemics'.) Of these, 631,000–827,000 were thought likely to have the ability to infect humans.[22] In addition, there are bacterial, fungal and other pathogens, which occur in a wide range of animal hosts.

The vast majority of these pathogens are highly unlikely to infect human hosts (Fig. 6.3). We may be exposed to many pathogens when we interact with animals, which virologists term as 'chatter'.[23] But there is a chance of infection in a human host from this chatter. Factors that predict the likelihood of human infection include the abundance of the existing donor host and the fraction of the host population that is infected, the frequency of human interactions with that host, and the probability of pathogen transfer during that encounter. The probability of a pathogen infection tends to decrease with increasing phylogenetic distance from the donor animal: if the donor animal is distantly related to humans with a very different physiology, the pathogens it harbours are less likely to be able to adjust to a human host. The probability of infection also varies between pathogens. For example, many trypanosomes have a wide host range, while others, such as the simian foamy viruses, have a narrow host range. The ability of the pathogen to evolve or adapt is another predictor of pathogen infection in humans.[24]

These factors can help explain why we have acquired pandemic-causing pathogens from animals. Chimpanzees have donated numerous zoonoses and diseases, including HIV/AIDS, because of their close phylogenetic relationship

Scientists believe that the outbreak of Ebola in 2013–16 began in the village of Méliandou, Guinea. *Saéz et al. (2014)*[19]

to us, and their physiology that largely matches ours. A high human abundance and movement of people facilitated the transport of HIV from chimpanzees in West Africa to cause an ongoing global disease pandemic. We have also gained pathogens from animals that are distantly related to humans, such as rodents, because of their high abundance and frequent association with humans. Many diseases have been acquired from domestic livestock, again, because of their abundance and frequent contact with humans. One example is measles, which probably evolved from the rinderpest virus in cattle. In contrast, many animals, including elephants, are not known to have donated pathogens to human hosts, probably because they are only distantly related, have infrequent encounters with humans and have a low abundance.[24]

The EIDs or zoonoses largely reflect our changing relationship with nature. The risk of virus transmission from donor animals has been highest from species that have increased in abundance and expanded their range due to adapting to human-dominated landscapes.[25] Domesticated species, primates, bats and rodents have been identified as harbouring high numbers of zoonotic viruses.

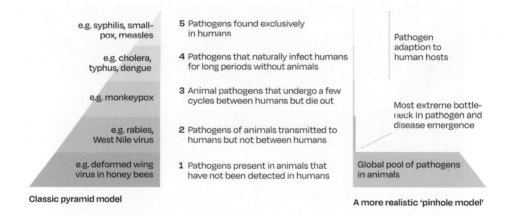

Fig. 6.3: The five stages through which pathogens from animals evolve to cause disease and become confined to humans. A large pool of pathogens infect animals. Humans are exposed to some of these through their interactions, and a small proportion are transmitted. Many are not transmissible between humans and require an animal vector (such as a mosquito, to transmit West Nile virus). Some pathogens exist for periods without animal transmission, with a smaller proportion found exclusively in humans. The classic pyramid model was proposed by Wolfe et al. (2007),[24] but the 'pinhole model' described by Warren and Sawyer (2019)[26] is probably more realistic, as the majority of pathogens in animals are unlikely to infect humans.

Of special interest are the bats, as they are central to both the 2013–16 Ebola outbreak and the ongoing COVID-19 pandemic. Bats do not display the strong activation of their immune system as a response to viral infection that we see in humans and other animals. Their immune response to viral infection appears adapted for pathogen tolerance rather than elimination. Consequently, they may host a wide diversity of viruses and typically exhibit asymptomatic infections from many of them.[27]

The exploitation of wildlife through hunting and trade has facilitated close contact between wildlife (including bats) and humans. The wet markets in Asia that sell a vast array of live and slaughtered animals, including bats, rodents and waterfowl, are great examples of such close contact. Bats are also used as food in many parts of Africa. This exploitation, combined with human activities that have resulted in reduced habitat quality, and an increased human population that has expanded our contact with animals, have increased opportunities for animal–human interactions. These interactions facilitate pathogen transmission and spillover to human populations.

The zoonotic RNA viruses that infect bats, waterfowl and many other

Friends and family look on as a body is carried away in a village in Sierra Leone, 2014. All bodies were to be buried within 24 hours in an authorised burial ground, by a trained Ebola burial team wearing full protective equipment. Some families felt that their right to show respect for the dead was being denied, which led to unsafe burials – a primary cause of Ebola infections. *Tommy Trenchard / Alamy*

animals have an amazing ability to change and adapt. These viruses have three characteristics that contribute to their ability to switch hosts and infect humans. They are tiny – you could fit 100 million SARS-CoV-2 virus particles on the head of a pin – and have massive populations. An infected organism might have 10 trillion virus particles. RNA viruses have short generations. A single infectious particle might produce 100,000 copies of itself in 10 hours, with a new RNA genome being produced every 0.4 seconds. And finally, RNA viruses have the highest mutation rates known of any living being, because they typically lack the cellular machinery for proofreading that you and I have. Those attributes create the perfect storm for viral population change and evolution.[28] While the vast majority of the mutants that are produced in each generation of viruses might die, a small number might have useful characteristics. These mutations, for example, might help it evade the immune system (antigenic drift). If multiple strains of an RNA virus occur in a host, we can also see 'antigenic shift', which is what happens in influenza viruses that combine to form a new viral strain with a mixture of surface proteins.

Emile Ouamouno, the first victim of Ebola virus in the 2013–16 outbreak, was an unfortunate recipient of RNA viral pathogen spillover or zoonosis. The altered landscape that probably increased the abundance of the Angolan free-tailed bats, and close contact between the children of Emile's village with the

An educational billboard in the village of Tuala, Sierra Leone, during the Ebola outbreak. *George Berg / Alamy*

tree where the bats roosted, increased the probability that viral diseases would spill over to the human population. The virus spread was aided by connectivity of the village to the wider region, which contained a substantial population. Without this connectivity, the Ebola virus might have killed a high proportion of the village but would not have dispersed to become an epidemic. This novel strain of the *Zaire ebolavirus* has been hypothesised to have adapted to human transmission and appeared to evolve over the course of the outbreak.[29]

## The spread of disease and pandemics

The simplest statistic that describes the spread of pathogens is the effective reproduction number of the pathogen, or $R_t$.

$R_t$ is the expected number of cases directly generated by each infected person in a population. A bit more simply, the $R_t$ number indicates the transmission potential, or the average number of people each person with a disease goes on to infect. In situations when $R_t$ is greater than 1, the pathogen and disease will spread to become epidemic, and if unchecked, potentially a pandemic. A disease with $R_t = 2.0$ means that one person will on average infect two others. When $R_t$ is sustained at values less than 1, the pathogen will stop spreading and will eventually disappear.

You might also hear people talk about the basic reproduction number, $R_0$ or $R$ nought, which is an estimate of the effective reproduction number of the pathogen when *all people are equally susceptible*. As a disease progresses through a population, some individuals recover to have immunity or are vaccinated, so different individuals will not be equal in their susceptibility. $R_t$ values are thus appropriate except right at the beginning of a disease outbreak and pathogen spread.[*]

Values of $R_t$ and $R_0$ vary substantially between diseases and environments,

---

[*] Epidemiologists worry that politicians and public alike place too much emphasis on $R_t$ values. A key problem is that $R_t$ is estimated retrospectively. Infections and diseases change, with local outbreaks in some areas but not others. That lag in time and space makes it difficult to use $R_t$ for future public health measures. In addition, different statistical models give different values. $R_t$ values are also averages and may ignore the presence of 'super-spreaders', who pass a disease to many more people than is typical. Another issue is zoonoses. With the bubonic plague, for example, we might lower the disease in humans, but it can still break out if the pathogen is prevalent in a huge population of rats and their fleas.[30]

and as a result of public health measures. The $R_0$ of measles, typically cited as 12–18 (one infected person will, on average, infect up to 18 others), represents one of the highest known values for any pathogen. In some populations, however, $R_0$ for measles appears to be much lower than these estimates, which has major implications for public health and management by herd immunity (Fig. 6.4). Populations displayed different values, dependent on many factors including the number of contacts, population density, birth rates and cultural practices.[31] Other diseases or pandemics have comparatively lower $R_0$ values, including MERS (0.29–0.80),[32] seasonal influenza (1.2–1.4),[33] Ebola (1.4–1.8)[34] and the 1918 flu (2.0–3.0).[35] The first detected SARS-CoV-2 virus had values ranging from 1.9–2.6,[36] although the following Delta variant was estimated to

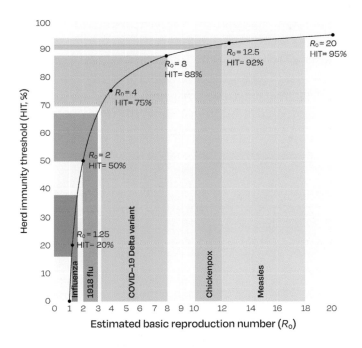

Fig. 6.4: The relationship between herd immunity threshold (HIT) and basic reproduction number ($R_0$) for selected diseases or pathogens. Note that the graph shows estimates, which vary between epidemiological models and with different data. $R_0$ values vary for each disease, depending on environment and human population. Of those shown, measles is the most infectious. This model suggests herd immunity would be reached for measles only if 92–95% of the population were immune, either by vaccination or previous infection. Influenza is much less infectious ($R_0$ = 1.2–1.4) and has an HIT of <40%.

be much higher at 3.2–8.0.[37] Pathogens evolve over time, typically to become more transmissible.

These basic reproduction numbers help explain why different viruses move through populations. The swift movement around the world of the SARS-CoV-2 virus Delta variant and its usurpation of previous variants is easily understandable. The SARS-CoV-2 Omicron variant then usurped Delta. New variants follow old as the viruses evolve.

What are the key reasons for such a wide variation in $R_t$ and variation in the spread of pathogens? Their spread is influenced by factors intrinsic to the pathogen. The physiology, demography and behaviour of the human host population also clearly influence epidemic and pandemic spread. For pathogens that require or use animals to vector or host the pathogen, the abundance and population dynamics of those animals is hugely important.

The key intrinsic factors that commonly promote pathogen transmission and high rates of spread include the ability to evade their hosts' immune system, the ability of the pathogen to attain a high abundance or high viral load, and having transmission at a low infectious dose.[38] The stability and survival of the pathogen on fomites, the external environment or surfaces external to the host, has been

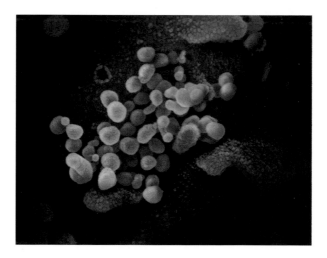

A scanning electron microscope image showing the SARS-CoV-2 virus. The virus particles were isolated from a patient in the US. Here they are emerging from the surface of cells (blue/pink) cultured in the laboratory. Colour has been added for definition. *National Institute of Allergy and Infectious Diseases*

found to be important for some pathogens. Different pathogens vary substantially in their use of hosts and the infections that result. For viruses, this is based on how and where they are able to enter or exit cells and co-opt the cells' machinery to reproduce themselves. Comparing the SARS-CoV family of pathogens and influenza, for example, shows that the coronaviruses have a longer incubation period (the time from infection to symptom onset) than influenza viruses.[35] These viruses also 'shed' or release infectious viruses differently. The initial SARS-CoV pathogen from 2003 had around a week between symptom onset and maximum virus loads and activity. This week-long delay could allow infected individuals to be identified and isolated before they infected others. Unfortunately, for the more recent SARS-CoV-2, symptom onset and maximum activity occur concurrently, making it harder to find and isolate sick people before they transmit the pathogen to others. (See 'Case study 3: A pandemic that stopped the world', page 250.)

We have learnt a lot about how diseases spread from the ongoing SARS-CoV-2 pandemic. Statistical models have consistently demonstrated patterns of disease spread that are highly correlated with human mobility. A nationwide programme to stop human movement stops the disease spread, even in large countries such as China.[39] In England, movement was found to be important, but so were a wide variety of demographic factors including the population density, ethnic and age demographics and the proportion of elderly (highly susceptible individuals), overcrowding factors including household density, and socio-economic factors including indices of deprivation.[40] We saw many of these factors influence the spread of the 1918 flu in New Zealand, when Māori communities were particularly affected. A wide range of other factors can influence pathogen spread, many of which are pathogen-dependent. Promiscuity, for example, is vital for HIV/AIDS to spread. With the Ebola virus, we've learnt that burial practices are an important method of transmission. Burial rites and customs, such as the ritual consumption of deceased relatives in Papua New Guinea, were necessary for the transmission of the prion disease Kuru (see Chapter 3, 'Case study 1: Prions, mad cows and sheep, and cannibalism', page 111). Once burial customs were changed, this prion disease went extinct.[41]

A key problem for managing diseases in human populations is that sometimes diseases can be 'hidden': pathogens can infect people completely asymptomatically. Perhaps the most famous asymptomatic carrier of a disease was Mary Mallon, who became known as Typhoid Mary. Mary was probably born with typhoid (the bacteria *Salmonella enterica* subsp. *enterica* serovar) after her mother became infected during pregnancy. She is not known to have experienced any symptoms, but passed the disease to at least 51 others, three of whom died.[42]

In the early 1900s scientists were beginning to suspect that some individuals could carry and spread disease without displaying symptoms themselves. A New York Department of Health official suspected that Mary, a cook, was an asymptomatic carrier of typhoid fever and approached her for an interview and, reportedly, a faecal sample. She allegedly threatened the investigator with a rolling pin or meat cleaver and chased him from her kitchen. Mary resisted the diagnosis till her dying day. She fought against her quarantine and isolation: 'Why should I be banished like a leper and compelled to live in solitary confinement with only a dog for a companion?' Of nearly 400 other asymptomatic carriers known at the time, Mary was the only one to be forced into an isolated quarantine for decades.

More recently, many SARS-CoV-2 virus carriers have been asymptomatic. Approximately 43% of cases in Iceland were asymptomatic and possibly 78% of cases in China had no idea they were infected and infectious.[35]

A second and related important problem for managing disease is the presence of 'superspreaders' and 'superspreading' events. Superspreaders are unusually contagious people or organisms who spread a disease at much higher than average rates. Similarly, superspreading events are those at which a disease is spread much more than is usual. Mary Mallon has been categorised as a superspreader of typhoid. We have known for over two decades that 20% of the host population contributes at least 80% of the disease transmission for both vector-borne parasites and sexually transmitted diseases.[43] Researchers now believe that superspreading is a normal feature in the spread of a wide variety of diseases. In fact, the most infectious individuals often cause substantially more than 80% of the infections for a wide variety of diseases including measles, SARS-CoV, HIV/AIDS and bacterial sexually transmitted diseases.[44] The presence of these superspreaders is tantalising for public health officials. If we could specially target control measures to the small portion of the population who are superspreaders, we could substantially slow disease spread.

## Elimination as a control method for pandemics

Controlling a pandemic is simple. All you need to do is lower and sustain the effective reproduction number of the pathogen, or $R_t$, to below 1. If you lower $R_t$

for an entire country for long enough, you'll achieve countrywide elimination or eradication. This simple control for pandemics is, however, much easier said than done. The methods used for disease control are classified into pharmaceutical and non-pharmaceutical means, with governments deciding on goals of suppression or the much harder elimination. The non-pharmaceutical means are sometimes referred to as public health and social measures.

The ideal goal for a pandemic is elimination. If a disease is gone entirely, then it will no longer have any negative health, economic or biodiversity effects. Complete eradication or global elimination has been achieved for only two viral pathogens and the diseases they cause: smallpox and rinderpest. Smallpox is a DNA virus thought to have infected the ancient Egyptian pharaoh Ramses V, who died in 1157 BC, and to have caused the 'Antonine Plague' in 165–180 AD (Fig. 6.1). It is thought to have evolved from a virus in rodents, with the advent of humans living in large settlements and farming livestock enabling its emergence.[45] Vaccination resulted in the elimination of the disease, which was declared globally eradicated in 1980. Rinderpest was a viral disease of cattle, buffalo and many other species of even-toed ungulates, including antelope and giraffes. It was similarly eradicated by mass animal vaccination in combination with zoosanitary procedures and adaptive management strategies that overcame unanticipated variations in viral pathogenicity or the circulation of disease in wildlife populations.[46] These two examples show how vaccines can be central to and effective for global disease elimination.

As with any proposed pest or disease mitigation, there is always debate on the best approach. Is elimination possible? Would the costs of elimination outweigh the benefits? These questions have been hotly debated during the global and New Zealand response to COVID-19, and countries around the world have shown very different responses in the management of this disease.

New Zealand quickly adopted a 'go hard, go early' elimination approach for COVID-19. A definition of 'elimination' for COVID-19 was proposed as: (1) the absence of newly diagnosed SARS-CoV-2 virus infections from community transmission, (2) the presence of a high-performing surveillance system and (3) the acceptance of suitable exemptions, such as cases of SARS-CoV-2 infection detected at the border among incoming travellers.[47] The New Zealand elimination strategy included border closure and management for exclusion and quarantine; high rates of rapid testing; and case- and contact-tracing using digital, genomic and genetic tools. Physical distancing (with the extreme of isolation by lockdown or stay-at-home orders), mask wearing and hand washing were also central to the New Zealand approach.[48] These non-pharmaceutical control methods quickly had a major effect, decreasing the effective reproduction number, $R_t$, from 7 to 0.2 within the first week of lockdown in 2020.[49] Nationwide elimination was subsequently achieved and declared within two months.

In the weeks after the COVID-19 'go hard and go early' plan was proposed, debate flared regarding the elimination plan for New Zealand. Those who opposed elimination questioned just how deadly COVID-19 actually is and gave statistics suggesting that it would cause similar mortality to influenza, that vaccines would take 3–4 years to develop, and that the costs of lockdown would vastly outweigh the benefits.[50] Hindsight is 20/20, but it is now clear that this initial elimination had significant benefits. Effective vaccines were developed at lightning speed. New Zealand's elimination approach resulted in the lowest cumulative COVID-19 death rate in the Organisation for Economic Co-operation and Development (OECD) with a rate 242 times lower than the average among the 38 OECD-member countries. Considering 'excess deaths', we experienced the largest negative value in the OECD, equivalent to around 2000 fewer deaths than would have been expected. Economically, in the average change in the gross domestic product indices, New Zealand was the sixth best performer. We also experienced less than average unemployment as a consequence of the pandemic.[51] However, there is no doubt the New Zealand response could be improved. We've still not learnt from the 1918 influenza pandemic, when Māori and Pacific peoples were disproportionately affected. From an analysis of 1829 COVID-19 cases in New Zealand, rates of hospitalisation were 2.5 times higher for Māori and 3 times higher for Pacific peoples than for other ethnicities.[52]

Some epidemiologists think it might be possible to globally eradicate COVID-19 with high vaccination rates.[53] They point to the recent elimination of malaria from China and the near complete global Guinea Worm Eradication Programme (neither of which had vaccines) as evidence of what can be achieved. After a US$16.5 billion investment, polio is nearing elimination as well, but it is sadly still endemic in Pakistan and war-torn Afghanistan. I suspect, however, that given wildlife reservoirs of COVID-19 discussed below, it will be a challenge to globally eradicate it.

## Disease suppression in pandemics

During the Great Plague of Marseille in 1720, attempts to stop the spread of the plague included an Act of the Parliament of Aix that could impose the death penalty for any communication between the people of Marseille and the rest of Provence. The Mur de la Peste, or Plague Wall, was erected to close the borders of the region. The wall and quarantine measures failed, unfortunately, as do many attempts at elimination or border restriction. Rather than aim for elimination, governments and civil authorities today might choose to suppress the disease and reduce its effects on a population and public health system.

Most of the non-pharmaceutical, public health and social measures that we use today to suppress pandemics have remained the same for many decades. For example, in the 1918 influenza pandemic, the interventions used in the United States included the closure of schools and churches, the banning of mass gatherings, mandated mask wearing, case isolation, and disinfection or hygiene measures. Cities in the United States implemented these interventions

The Great Plague of Marseille was the last major outbreak of bubonic plague in western Europe, killing a total of 100,000 people. A plague wall was erected in an attempt to enforce their isolation. This boundary marker is a remnant. *Hemis / Alamy*

differently; those that implemented them earlier and maintained them for longer experienced lower rates of mortality.[54] Different towns and regions in New Zealand also experienced substantial variation in mortality rates. Historian Geoffrey Rice wrote, 'The Grim Reaper of the 1918 influenza pandemic found an abundant but uneven harvest in New Zealand, as if the seeds had been scattered by a drunken gardener: here a thick handful, there a thin sprinkling, and in some places hardly any at all'.[11] Wellington stood out with the highest death rate of 7.9 per thousand, well above the national average; Nelson had a low death rate, of 2.2 per thousand. Differences in population density and in the use of public health and social measures helped explain the variation in death rates between cities. In more recent times, border closures, quarantine, lockdown, case- and contact-tracing and other approaches described above have been the critical non-pharmaceutical tools in New Zealand's effective approach to COVID-19.

*The Cow Pock — or — the Wonderful Effects of the New Inoculation!* — Vide the Publications of ye Anti-Vaccine Society.

The world's first vaccine is credited to British physician Edward Jenner. Because of his vaccine against cowpox, he has been described as having saved more lives than any other human. A cartoon titled 'The Cow-Pock – or – the Wonderful Effects of the New Inoculation!' (1802) portrayed people's fears of vaccination, such as that it would make them sprout cow-like appendages. The text reads 'Vide – the Publications of ye Anti-Vaccine Society'. Anti-vaccination groups have been with us right from the beginning. *British Cartoon Prints Collection*

Vaccines are the key pharmaceutical tools in minimising the health effects of pandemic diseases. Vaccines can provide individual recipients with a high probability of complete protection from a disease, or protection from serious illness if they do become infected. Vaccine efficacy is typically reported as a 'relative risk reduction', which estimates the ratio of attack rates with and without a vaccine. For the COVID-19 vaccines, the relative risk reductions have been estimated to range from 95% for the Pfizer–BioNTech to 67% for the Oxford/AstraZeneca vaccines.[55] For comparison, the seasonal influenza vaccine effectiveness is much lower at 45% in 2019–20,[56] and only approximately two-thirds of eligible New Zealanders receive the vaccine. That combination – relatively low efficacy and a small percentage of the population being vaccinated – sounds like a problem, but depending on the strain-specific $R_t$ value, it is likely to have been enough to provide a reasonable level of herd immunity for the influenza A virus (as discussed below; Fig. 6.4). Changes or mutations in influenza viruses caused by antigenic drift mean that new influenza shots are needed each year.

The protection offered by vaccines often declines over time. Research on the Delta strain of SARS-CoV-2 has shown the effectiveness of the Pfizer–BioNTech vaccine to decline quickly to 90%, 85% and 78% after 30, 60 and 90 days, respectively.[57] We would expect it to drop further after more time. Other COVID-19 vaccines have shown similar declines, indicating booster shots or entirely new vaccines may be needed for this or newer strains of this virus. With diseases such as measles, for comparison, a single vaccination can provide more than 96% efficacy and offer lifelong protection. These differences in vaccine efficacy and delivery are due to many factors, including how the vaccine is prepared. Measles vaccines, for example, are made from attenuated measles viruses, which have been substantially weakened in the laboratory production process. It is thought that using an entire pathogen allows your immune system to develop a memory of many different targets of the virus. Using just a spike protein target, such as with the SARS-CoV-2 Pfizer–BioNTech vaccine, results in a less diverse immune memory. The hunt is on for a more effective and general, broad-spectrum, universal coronavirus vaccine.[58]

Herd immunity is frequently promoted with the use of vaccines and can occur through vaccination or previous infections. There is a simple predicted relationship between the basic reproduction number $R_0$ and herd immunity. As $R_0$ increases, a larger and larger proportion of the population is predicted to require vaccination to achieve herd immunity (Fig 6.4). Diseases such as influenza have relatively low rates of infectivity and require a smaller proportion of the population to be vaccinated to achieve herd immunity. Because measles and chickenpox viruses are so infectious, you need to have more than 90% of the population either vaccinated or having developed immunity to these diseases to achieve herd immunity. Vaccination programmes can be effective. In Japan during the 1970s and 1980s, legislation made vaccination against influenza compulsory for school children. Although the elderly were the high-risk group, school children were targeted because they were known to spread and amplify the disease. Their vaccination was estimated to have prevented 37,000–49,000 deaths each year, or approximately one death for every 420 children vaccinated.[59]

Like nearly every beautifully simple theory, the relationship between herd immunity and $R_0$ has proven to be more complex than at first glance. These complexities arise because of imperfect immunity, heterogeneous populations, non-random vaccination and 'freeloaders'.

Imperfect immunity arises when vaccination or even prior illness does not confer complete immunity against infection to all recipients. They may still catch the disease in the future with a mild or more serious illness, but importantly they can still spread the pathogen. Unfortunately, imperfect immunity appears to be happening with COVID-19 vaccinations. Heterogeneous populations are those with different groups, such as a variety of age structures or other demographics, that mix and mingle. Those differences mean that $R_0$ or $R_t$ values can differ substantially between groups. Variation between groups also results in patterns of non-random vaccination and disease susceptibility. Some religious groups or sectors of society shun or ban vaccination, and when they congregate, elevate infection risk. Similarly, social clustering frequently occurs among parents who decide not to vaccinate their children, resulting in groups of children and parents in which vaccination rates are considerably below the threshold for herd immunity. Finally, the proportion of 'freeloaders' in the population can also influence herd immunity. Vaccination will have some cost to each individual. Side effects, costs in money or time, or inconvenience can influence a person's decisions on vaccination. Individuals who consider the costs too high might decide to rely on everyone else to become vaccinated, and if costs are too high for too many, these freeloading individuals may have a major impact on achieving the levels of vaccination required for herd immunity.[60]

These complexities associated with vaccination and herd immunity are not just theoretical. Ben Adler from the Department of Microbiology at Monash University wrote in 2021 about how vaccination responses paint a worrying picture in Australia:

> If the figures of 11 per cent of the population refusing COVID-19 vaccination outright and 24 per cent hesitant to do so are correct, we are in real trouble. We will never achieve herd immunity against the virus.
>
> It is therefore time to call out these people for what they really are: bludgers and freeloaders who expect the rest of us to take the very small risk and very small discomfort associated with vaccination in order to limit the spread of virus in the community, thereby protecting that community, which of course includes those who are not vaccinated.[61]

## Governmental legislation for pandemic management in New Zealand

The WHO plays a major global role in the response to potential pandemics, using notifications of disease that may constitute a Public Health Emergency of International Concern (PHEIC). As discussed in Chapter 2, WHO member states use an algorithm to determine if a PHEIC exists, based on any two of the four following questions:

Is the public health impact of the event serious?

Is the event unusual or unexpected?

Is there a significant risk of international spread?

Is there a significant risk of international travel or trade restrictions?

Member states are obligated to respond to PHEIC events. The New Zealand response to a pandemic involves three pieces of legislation: the Health Act 1956, the Civil Defence Emergency Management Act 2002 (CDEM Act 2002) and the Epidemic Preparedness Act 2006.

The Health Act 1956 is the key statute for the containment of communicable diseases at the border and within the country. This act gives powers to the director-general of the Ministry of Health, the minister of health and medical officers to respond to outbreaks of diseases that are specified in a schedule of the Health Act. The 'novel coronavirus capable of causing severe respiratory illness' (SARS-CoV-2 and COVID-19) was listed on the schedule as a notifiable infectious disease in January 2020. The use of this act enables the government to detain and isolate individuals that are a public health risk; to 'require persons, places, buildings, ships, vehicles, aircraft, animals, or things to be isolated, quarantined, or disinfected'. It enables the closure of premises and authorises contact-tracing.[62] The Health Act 1956 works alongside the more general CDEM Act 2002, which enables road closures or traffic regulation. The civil defence actions can go further to enable government representatives to evacuate places or even enforce specific methods to dispose of the dead.[63]

The Epidemic Preparedness Act 2006 was developed to provide additional legislation, the need for which was highlighted by emerging diseases such as SARS-CoV and the threat of influenza A (H5N1). An 'epidemic notice' is issued when the prime minister is satisfied that the effects of an outbreak of a quarantinable disease (as defined in the Health Act 1956) are likely to disrupt essential governmental and business activity in New Zealand. It enables health officials to enact all of the special powers set out in sections 70 and 71 of the Health Act 1956, including having people submit to medical examinations. The government can control the prescription and supply of medicine or medical equipment. Under an epidemic management notice, the Ministry of Social Development may grant emergency benefits to people who would otherwise not be entitled to such benefits.[64]

New Zealand's response to the COVID-19 pandemic has been shaped by the government's 'Influenza Pandemic Plan'. We have had a such a plan since 2006, but it was updated in 2017 when the next pandemic was thought most likely to be from an influenza A virus. The plan works on six potential pandemic phases: 'plan for it', 'keep it out', 'stamp it out', 'manage it', 'manage it (post-peak)' and 'recover from it'.[65] New Zealand's initial 'go hard and go early' action with COVID-19 fell under the 'stamp it out' phase. In early 2022 we moved to a 'manage it' response once the vast majority of the population were vaccinated. A 'recover from it' phase will be lengthy.

## Plant or animal pandemics and reverse zoonoses

Animals and plants experience pandemics too, and although they often receive less publicity, their pandemics can be just as devastating.

A recent pandemic in plant communities involves the fungal pathogen myrtle rust (*Austropuccinia psidii*), native to South America. Over the last three decades, a new 'pandemic' strain of this pathogen emerged and has caused severe epiphytotics (the equivalent of epidemics in plants) in Hawai'i, Japan, China, Australia, South Africa and New Zealand.[66] High levels of infection can kill entire stands of trees in less than four years, resulting in localised extinction. It affects plants in the family Myrtaceae, giving the leaves or shoots a distinctive yellow or orange colour.

After its arrival myrtle rust was cited as 'probably the biggest threat to Australia's ecosystem'.[67] Almost 80% of the trees in Australia belong to the family Myrtaceae, including the iconic gum trees. One Australian botanist said, 'From a plant perspective I don't think any (invasive plant pathogen) compares to myrtle rust because it spreads so quickly and affects so many different plant species, many of which are key cornerstone species critical to fragile environments. This has been the pinnacle of pathogens we wanted to keep out of Australia.'[67] It is not known how the fungus slipped into Australia, where initial plans to eradicate it were soon abandoned. Australia's quarantine incursion and failure to eradicate the pest have been bad news for New Zealand. Myrtle rust was first observed here in 2017, after wind-blown spores from Australia came to call New Zealand home.

A currently ongoing and worsening pandemic in animals is the deformed wing virus in honey bees (*Apis mellifera*), spread by the parasitic mite *Varroa destructor*.[68] The mite and the virus have spread from their native host, the Asian honey bee (*Apis cerana*), to the evolutionary naïve European honey bee (*Apis*

Orange myrtle rust (*Austropuccinia psidii*) on lilly pilly (*Syzygium smithii*) leaves. *SmallBiologie / Wikimedia Commons*

*mellifera*) over the last century, to become 'the greatest threat to honey bee health'.[69] The strains of virulent and recombinant forms of the deformed wing virus are closely linked with honey bee losses.[70] This is a human-made pandemic in a globally important pollinator, propelled by our inadvertent spread of the parasitic mite.

Animals have gifted us many of the diseases causing pandemics in humans. 'Reverse zoonosis' or 'zooanthroponosis' is when humans return these diseases to animal populations. A wide array of bacterial, viral, fungal and many other pathogens or parasites have been observed to flow back to animals from humans[71] on every continent, including Antarctica.[72] We've seen and learnt a lot about these zooanthroponosis events in two of the more recent pandemics: the H1N1 influenza virus in 2009 and SARS-CoV-2 that causes COVID-19 disease. A wide variety of farmed, pet and wild animal populations were infected by these pathogens. Farmed animals including turkeys, pigs and mink were infected and died from these diseases.[73] Both pathogens were seen in domestic cats, and in

A worker bee with stunted and deformed wings due to the deformed wing virus, compared with a healthy bee worker. Parasitic *Varroa destructor* mites, shown on the deformed bee, transmit this virus. When the mites fed on the pupae, the virus infected the developing bee. The virus then multiplied in the developing wing buds, causing deformities. The worker also has fewer hairs and cannot fly. She will be ejected from the hive by fellow workers and will die.
*Phil Lester photo*

cheetahs and lions in zoos. Wild populations of skunks were infected with the H1N1 influenza virus, and a substantial proportion of wild deer in the United States now carry SARS-CoV-2 infections.[74]

Several conclusions can be reached from the zoonoses in these populations. Animals that interact closely with humans are most at risk. It is thought that turkey populations became infected with the H1N1 influenza virus during close human contact in artificial insemination procedures. For these viruses, attaining a high viral load and shedding rates are important. Intraspecific contact in the recipient animal population is crucial for the virus to persist, so animals living in groups or herds and in high densities pose a high risk.

The methods that can be used to control or eradicate a pandemic in animal or plant populations are largely similar to those discussed for humans. One of the two animal pathogens globally eliminated, the rinderpest virus, was largely overcome as a result of animal vaccination. In New Zealand we have seen the countrywide elimination of several widespread pathogens. Hydatid disease, for example, is caused by the tapeworm *Echinococcus granulosus*, a parasite that lives in the small intestine of dogs as an adult, but which has intermediate hosts that can include sheep and humans. This parasite has a global distribution that included New Zealand, where the worm was found in nearly all mobs of sheep as early as 1897. The government implemented control via the Hydatids Act 1959, which employed a range of pharmaceutical and non-pharmaceutical control methods, including a dog-owner education campaign.[75] Dog registration and de-worming treatments became compulsory. Stock vaccination was attempted but abandoned. The last observation of hydatid disease in New Zealand was in 1996; the tapeworm is now considered eradicated.[76]

The eradication of diseases such as hydatids demonstrates an ability to control pathogens. On a less positive note, the solutions or controls for the ongoing multi-host plant and animal pandemics, including myrtle rust, represent a considerable challenge.

## A future with more frequent, severe pandemics

At the start of this chapter I described how we are in a pandemic era. The bad news is that there is no light at the end of this tunnel. Marco Marani and colleagues recently published a major study examining the intensity and frequency of extreme novel epidemics. Their work concludes that a future with extreme epidemics is likely, and 'the yearly probability of occurrence of extreme epidemics can increase up to threefold in the coming decades'.[77] Just as our current COVID-19 pandemic is of zoonotic origin, it seems highly likely that our next epidemic and pandemic will also be an RNA virus with an animal origin. Our global connectivity and highly efficient transport systems mean that this 'Disease X' will have the ability to move to almost anywhere in the world within a day or two.

There is a strong consensus that we need to switch from a reactive to a proactive role in dealing with pandemics. The One Health initiative that recognises the interconnections between human and veterinary medicine and their shared environment may be one way to achieve this goal. There are examples of such proactive programmes already. For example, the United States Agency for International Development PREDICT project examines the zoonotic spillover of viruses in 31 countries. PREDICT has found and described over 800 novel viruses, but has also used community education programmes that raise awareness of zoonotic disease risks, with a relatively small budget of US$200 million over 10 years.[78] Cost–benefit analyses of these types of programmes, using both surveillance and proactive measures, are hugely in favour of proactive approaches. One analysis suggested the global total gross prevention costs for pandemics would range between US$22 and 31 billion. These costs include programmes for early detection and control, reducing spillover from livestock, reducing deforestation and ending the wild meat trade in China.

In comparison, the disease damage from the COVID-19 pandemic alone has been estimated at between US$8.1 and 15.8 trillion.[78]

## Case study 1: Finding and reconstructing the H1N1 'Spanish flu' pandemic virus

The 1918 H1N1 flu pandemic was caused by an influenza A virus. These viruses are common in waterfowl and shorebirds. More than 100 of these birds host a wide diversity of influenza A viruses, with infections typically appearing to have little impact. They represent a global generator and reservoir of new flu strains, although the vast majority are not infectious in people. Occasionally, however, an influenza A virus arises that can infect human cells (and other animals) that can cause a pandemic. Of all the strains, the 1918 H1N1 virus has so far proved the most lethal. But where did it originate, and from what host? Why was it so lethal?

The key to answering questions about the origin and effects of the 1918 H1N1 flu pandemic virus is its genetic composition. Describing the genome of this pathogen has for decades been the goal of virus hunters, who made some progress sampling formalin-preserved lung tissue from victims. But these samples were degraded; better-preserved samples were desperately needed. A major breakthrough came in the work of Johan Hultin, a retired Swedish-American pathologist who has been described as 'part Indiana Jones, part Anthony Fauci'.[79] Johan knew the location of a graveyard in Alaska where victims had been frozen in the permafrost of the village of Brevig Mission and were still well preserved. Financing the expedition himself, he flew to Brevig Mission. The 1918 pandemic had claimed 72 of the 80 inhabitants. The council allowed the grave excavation, and Johan uncovered an Inuit woman in her mid-20s who had died from complications associated with the 1918 flu. Her lungs were to yield viral fragments that, when combined with sequence data from two US servicemen, enabled the reconstruction of the sequences for the surface proteins of the 1918 H1N1 flu pandemic. Surface proteins are key: they enable the virus to bind to

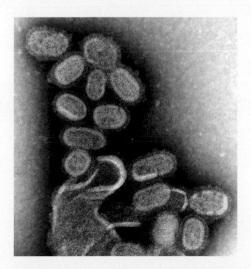

A coloured transmission electron microscope image of the virus responsible for the 1918 flu pandemic, sometimes called Spanish flu. *Cynthia Goldsmith / CDC*

and enter and infect cells. A series of publications went on to describe all eight gene sequences, effectively the entire genome of the 1918 H1N1 flu virus.[10]

A phylogenetic analysis indicated that the 1918 pandemic had an avian origin, but it had been circulating in and adapting to an unknown mammalian host before arising as a human pandemic. Some of the earliest known cases were in the US from military camps, not from Spain.[10]

In order to understand the pathogenicity of the virus, scientists used the genome to reconstruct the 1918 H1N1 flu pandemic virus in a US laboratory in 2005. This was a risky procedure and security was paramount, involving iris and fingerprint scans for the sole scientist performing the work.[80] Experiments showed that the virus was highly infectious and replicated to much higher levels in mice. It was also highly lethal. The lung tissue of mice was severely damaged, and pneumonia was common. It appears that this particular genetic recombination and evolution made the virus devastating. The researchers wrote that 'the constellation of all eight genes together make an exceptionally virulent virus' and 'no other human influenza viruses that have been tested show a similar

Johan Hultin at the Brevig Mission gravesite in Alaska, 1997, recovering lung tissue from the body of an Inuit woman who died from 1918 flu caused by the H1N1 influenza A virus. Because of the permafrost the woman's lungs were preserved. *Johan Hultin / CDC*

pathogenicity'.[81] To date there have been more influenza A pandemics, but none have been as virulent. The descendants of this 1918 H1N1 flu pandemic virus are still circulating in animal and human populations, but in a less lethal form.

## Case study 2: The 2020 IPBES report on biodiversity and pandemics

A report on biodiversity and pandemics from the Intergovernmental Platform on Biodiversity and Ecosystem Services (IPBES) was released in 2020.[22] The report came from a workshop that brought together experts from around the world to discuss: 1) how pandemics emerge from the microbial diversity found in nature; 2) the role of land-use change and climate change in pandemics; 3) the role of wildlife trade in driving pandemics; 4) learning from nature to better control pandemics; and 5) preventing pandemics based on a 'One Health' approach. Here I summarise the major findings taken from their report.

Pandemics represent a major threat to our health and welfare. They are becoming more frequent, driven by a continuing rise in underlying emerging disease events. Without preventative strategies, pandemics will emerge more often, spread more rapidly, kill more people and affect the global economy with more devastating impact than ever before. The recent exponential rise in consumption and trade, driven by demand from developed countries and emerging economies as well as by demographic pressure, has led to a series of emerging diseases that originate mainly in biodiverse developing countries and are driven by global consumption patterns.

*1. Pandemics emerge from the microbial diversity found in nature.* The majority (70%) of emerging diseases (such as Ebola, Zika, Nipah encephalitis), and almost all known pandemics (such as influenza, HIV/AIDS, COVID-19), are zoonoses – they are caused by microbes of animal origin. These microbes 'spill over' due to contact among wildlife, livestock and people. An estimated 1.7 million currently undiscovered viruses are thought to exist in mammal and avian hosts. Of these, 631,000–827,000 could have the ability to infect humans. The most important reservoirs of pathogens with pandemic potential are mammals (in particular bats, rodents, primates) and some birds (in particular water birds), as well as livestock (such as pigs, camels, poultry).

*2. Human ecological disruption and unsustainable consumption drive pandemic risk.* The risk of pandemics is increasing rapidly, with more than five new diseases emerging in people every year, any one of which has the potential to become pandemic. The risk of a pandemic is driven by exponentially increasing anthropogenic changes. Blaming wildlife for the emergence of diseases is thus erroneous, because emergence is caused by human activities and the impacts of these activities on the environment. Unsustainable exploitation of the environment – due to land-use change, agricultural expansion and intensification, wildlife trade and consumption and other drivers – disrupts natural interactions among wildlife and their microbes, increases contact among wildlife, livestock, people, and their pathogens and has led to almost all pandemics. Climate change has been implicated in disease emergence.

*3. Reducing anthropogenic global environmental change may reduce pandemic risk.* Pandemics and other emerging zoonoses cause widespread human suffering, and likely more than a trillion dollars in economic damages annually. Conservation of protected areas, and measures that reduce unsustainable exploitation of high biodiversity regions, will reduce the wildlife–livestock– human contact interface and help prevent the spillover of novel pathogens.

*4. Land-use change, agricultural expansion and urbanisation cause more than 30% of emerging disease events.* Land-use change is a globally significant driver of pandemics and has caused the emergence of more than 30% of new diseases reported since 1960. Land-use change creates synergistic effects with climate change (forest loss, urban heat island effects, burning of forest to clear land) and biodiversity loss that in turn have led to important emerging diseases. Destruction of habitat and the encroachment of humans and livestock into biodiverse habitats provide new pathways for pathogens to spill over and increase transmission rates.

*5. The trade and consumption of wildlife is a globally important risk for future pandemics.* Wildlife trade has occurred throughout human history. It provides nutrition and welfare for peoples, especially Indigenous peoples and local communities. The farming, trade and consumption of wildlife and wildlife-derived products (for food, medicine, fur and other products) have led to biodiversity loss and emerging diseases, including SARS and COVID-19.

*6. Current pandemic preparedness strategies aim to control diseases after they emerge. These strategies often rely on, and can affect, biodiversity.* Our business-as-usual approach to pandemics is based on containment and control after a disease has emerged. It relies primarily on reductionist approaches to vaccine and therapeutic development, rather than on reducing the drivers of pandemic risk to prevent them before they emerge. Pandemic control programmes often act under emergency measures and can have significant negative implications for biodiversity, such as the culling of wildlife reservoirs and the release of insecticides.

*7. Escape from the pandemic era requires policy options that foster transformative change towards preventing pandemics.* The major impact that COVID-19, HIV/ AIDS, Ebola, Zika, influenza, SARS and many other emerging diseases have on public health underlines the critical need for policies that promote pandemic prevention, based on this growing knowledge. To achieve this, several policy options were identified. These policy options include launching a high-level intergovernmental council on pandemic prevention, and implementing the One Health approach (that is, one that links human health, animal health and environmental sectors). New policies are needed to reduce the role of land-use

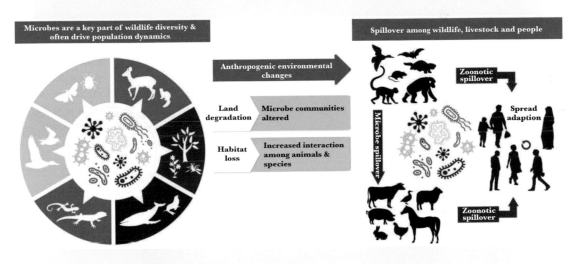

Fig 6.5: The origins and drivers of emerging zoonotic diseases and pandemics. Microbes have evolved within species of wildlife over evolutionary time. They undergo complex life cycles of transmission among host species, and often have significant impacts on host population dynamics. These microbes become emerging infectious diseases (EIDs) when anthropogenic environmental changes alter the population structure of their reservoir hosts, and bring wildlife, livestock and people into contact. These interactions can alter transmission dynamics of microbes within their hosts, lead to interspecies transmission of microbes, spill over to livestock and people, and lead to the emergence of novel diseases. While many outbreaks are small scale or regional, some EIDs become pandemics when zoonotic pathogens transmit easily among people, and spread in rapidly urbanising landscapes, megacities and travel and trade networks. Pandemics are a subset of EIDs. *IPBES (2020)²²*

hange in pandemic emergence, and to reduce pandemic emergence related to the wildlife trade. There is a need to foster a role for all sectors of society to engage in reducing the risk of pandemics.

A key conclusion from this report is that we need to change from a reactive to a proactive role in dealing with pandemics. Clearly, in the face of COVID-19, with more than one million human deaths and huge economic impacts, our current reactive approach is inadequate.

## Case study 3: A pandemic that stopped the world

The SARS-CoV-2 pathogen is the cause of the COVID-19 disease. The virus name is short for the 'severe acute respiratory syndrome coronavirus'. In case you were wondering, the '-2' on the end tells us that we've previously seen a related virus. This original SARS-CoV also first appeared in China. It spread to 29 countries, infecting approximately 8000 people in 2003.[82] Horseshoe bats (*Rhinolophus* spp.) in China were found to harbour related coronaviruses and it is thought that they passed it to humans via an intermediate host, probably palm civets (*Paguma larvata*).[83] A captive civet cat probably became infected after direct contact with bats in 'wet markets', where animals including horseshoe

A caged Asian palm civet (*Paguma larvata*). These animals were a likely intermediate host for the SARS-CoV outbreak in 2003. They are used for the production of Kopi Luwak, a coffee made from cherries that the animal has eaten and defecated. *Panther Media GmbH / Alamy*

bats and civets are sold alive or butchered.

Horseshoe bats also host viruses very similar to SARS-CoV-2, which first emerged in Wuhan city in China. We don't know for sure how the virus first infected humans or if there was an intermediate host. A virology laboratory exists in Wuhan and it is possible that the pathogen escaped or leaked from this lab. Some virologists believe the lab-leak hypothesis to be 'improbable', while others see potential 'smoking gun' evidence of human tampering in the genes associated with furin cleavage site on the virus's important spike protein.[84] While we don't have enough evidence to rule out a lab-leak hypothesis, it is also true that a wide variety of live animals were for sale in Wuhan markets, including civet cats, foxes, minks and raccoon dogs, all of which are susceptible to coronavirus. Monthly sales at this market comprised approximately 36,295 individual animals belonging to 38 terrestrial wild animal species. Of these, 31 were protected species. Nearly all animals were kept in stacked cages and in poor condition, often with wounds from capture or hunting. They were sold alive, though most sellers offered butchering services on site. No bats were sold at the market.[85]

Two horseshoe bats (*Rhinolophus sp.*). These bats occur in Europe, Northern Africa, Central Asia and Eastern Asia.
*Rudmer Zwerver / Alamy*

Scientists believe that SARS-CoV-2 is a highly generalist virus.[82] The virus uses a protein on the surface of cells to gain entry into humans: the angiotensin-converting enzyme 2 (or ACE2). SARS-CoV-2 is capable of using the ACE2 proteins of a wide variety of animal host species. Dogs, cats and many other animals have ACE2 proteins that are highly permissive for viral attachment.[86] We now see SARS-CoV-2 in animals including domesticated cats, dogs and ferrets, as well as captive-managed mink, lions, tigers, deer and mice. One study in Michigan found 67% of white-tailed deer were infected.[74] Humans have spread this disease to these animals. Interestingly, bats were poor hosts for the virus, which seems to indicate that the SARS-CoV-2 receptor has changed during zoonotic transmission and evolution from bats to people.[86]

In humans ACE2 is unfortunately common in lung tissue, with expression substantially increased in the lungs of patients with comorbidities such as hypertension or chronic obstructive pulmonary disease.[87] This increased ACE2 expression helps explain why COVID-19 is such a problem for people who have comorbidities. The virus is continuing to evolve and become more virulent.

The Delta variant made SARS-CoV-2 approximately 40% more transmissible. How? The furin cleavage site on the virus's spike protein again appears to have played a role. A single amino acid change means that the virus fused with human cells almost three times faster than before.[88] The Omicron variant had even more mutations on this spike protein. The SARS-CoV-2 pathogen will continue to change, and we will continue to see its evolution.

## Further reading and discussion

1. Most of the viruses that cause human disease come from other animals. This has led epidemiologists to attempt 'zoonotic risk prediction' in an attempt to forecast the next pandemic virus. Other scientists say this approach is of limited value, and that instead we should focus on the human–animal interface for intensive viral surveillance. Is such prediction or surveillance useful? Are the two approaches distinct? How would you suggest we predict our next pathogen?

*See:* Wille et al. (2021). How accurately can we assess zoonotic risk? *PLoS Biology, 19*(4), e3001135. doi.org/10.1371/journal.pbio.3001135

2. A survey from the Classification Office of the New Zealand government offered a glimpse at how pervasive misinformation is in New Zealand. The survey showed one in two Kiwis held at least one belief based on misinformation. These ranged from the belief that scientists are lying about the safety of vaccines, to the belief that 5G communications cause Covid-19. Are these beliefs a problem for pandemics? And if so, how can we solve the misinformation problem?

*See:* Classification Office (2021). *The edge of the infodemic: Challenging Misinformation in Aotearoa.* classificationoffice.govt.nz/assets/PDFs/Classification-Office-Edge-of-the-Infodemic-Report.pdf

3. 'One Health' principles have been suggested to be useful towards preventing globally widespread pandemics that result from zoonoses. What are the key principles, and how might we use them for preventing future pandemics? If we were to implement these measures in New Zealand, how would we do so?

*See:* Bird, B. H. & Mazet, J. A. K. (2018). Detection of emerging zoonotic pathogens: An integrated One Health approach. *Annual Review of Animal Biosciences, 6,* 121–139. doi.org/10.1146/annurev-animal-030117-014628

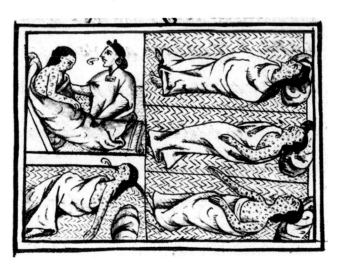

Illustration from the sixteenth-century Florentine Codex, compiled by the Franciscan friar Bernardino de Sahagún, showing the effects of smallpox on Indigenous populations.
*Wikimedia Commons*

# References

1.  The Lancet Planetary Health. (2021). A pandemic era. *The Lancet Planet Health, 5*(1), e1. doi.org/10.1016/S2542-5196(20)30305-3

2.  Demeure, C. E., Dussurget, O., Mas Fiol, G., et al. (2019). *Yersinia pestis* and plague: An updated view on evolution, virulence determinants, immune subversion, vaccination, and diagnostics. *Genes & Immunity, 20*(5), 357–370. doi.org/10.1038/s41435-019-0065-0

3.  Spyrou, M. A., Tukhbatova, R. I., Wang, C. C., et al. (2018). Analysis of 3800-year-old *Yersinia pestis* genomes suggests Bronze Age origin for bubonic plague. *Nature Communications, 9*, Article 2234. doi.org/10.1038/s41467-018-04550-9

4.  Zimbler, D. L., Schroeder, J. A., Eddy, J. L., et al. (2015). Early emergence of *Yersinia pestis* as a severe respiratory pathogen. *Nature Communications, 6*, Article 7487. doi. org/10.1038/ncomms8487

5.  Cohn, S. K. (2002). The black death: End of a paradigm. *The American Historical Review, 107*(3), 703–738. doi.org/10.1086/ahr/107.3.703

6.  Gold, H. (2019). *Japan's infamous Unit 731: Firsthand accounts of Japan's wartime human experimentation program.* Tuttle Publishing.

7.  Valles, X., Stenseth, N. C., Demeure, C., et al. (2020). Human plague: An old scourge that needs new answers. *PLoS Neglected Tropical Diseases, 14*(8), Article e0008251. doi.org/10.1371/journal.pntd.0008251

8.  Pauli, J. N., Buskirk, S. W., Williams, E. S., et al. (2006). A plague epizootic in the black-tailed prairie dog (*Cynomys ludovicianus*). *Journal of Wildlife Diseases, 42*(1), 74–80. doi.org/10.7589/0090-3558-42.1.74

9.  Potter, C. W. (2001). A history of influenza. *Journal of Applied Microbiology, 91*(4), 572–579. doi.org/10.1046/j.1365-2672.2001.01492.x

10. Taubenberger, J. K., Kash, J. C., & Morens, D. M. (2019). The 1918 influenza pandemic: 100 years of questions answered and unanswered. *Science Translational Medicine, 11*(502), Article eaau5485. doi.org/10.1126/scitranslmed.aau5485

11. Rice, G. W. (2005). *Black November: The 1918 influenza pandemic in New Zealand.* Canterbury University Press.

12. Sender, R., Fuchs, S., & Milo, R. (2016). Revised estimates for the number of human and bacteria cells in the body. *PLoS Biology, 14*(8), Article e1002533. doi.org/10.1371/journal.pbio.1002533

13. Balloux, F., & van Dorp, L. (2017). Q&A: What are pathogens, and what have they done to and for us? *BMC Biology, 15*, Article 91. doi.org/10.1186/s12915-017-0433-z

14. Woolhouse, M. & Gaunt, E. (2007). Ecological origins of novel human pathogens. *Critical Reviews in Microbiology, 33*(4), 231–242. doi.org/10.1080/10408410701647560

15. Smith, K. F., & Guégan, J.-F. (2010). Changing geographic distributions of human pathogens. *Annual Review of Ecology, Evolution, and Systematics, 41*(1), 231–250. doi. org/10.1146/annurev-ecolsys-102209-144634

16 Faruque, S. M., Islam M. J., Ahmad, Q. S., et al. (2005, April 13). Self-limiting nature of seasonal cholera epidemics: Role of host-mediated amplification of phage. *Proceedings of the National Academy of Sciences, 102(17)*, 6119–6124. doi.org/10.1073/pnas.0502069102

17. Haider, N., Rothman-Ostrow, P., Osman, A. Y., et al. (2020). COVID-19 - Zoonosis or emerging infectious disease? *Frontiers in Public Health, 8*, Article 596944. doi.org/10.3389/fpubh.2020.596944

18. Roychoudhury, S., Das, A., Sengupta, P., et al. (2020). Viral pandemics of the last four decades: Pathophysiology, health impacts and perspectives. *International Journal of Environmental Research and Public Health, 17*(24), Article 9411. doi.org/10.3390/ijerph17249411

19. Mari Saéz, A., Weiss, S., Nowak, K., et al. (2015). Investigating the zoonotic origin of the West African Ebola epidemic. *EMBO Molecular Medicine, 7*(1), 17–23. doi.org/10.15252/emmm.201404792

20. Kupferschmidt, K. (2021). Ebola virus may lurk in survivors for many years. *Science, 371*(6535), 1188. doi.org/10.1126/science.371.6535.1188

21. Anthony, S. J., Epstein, J. H., Murray, K. A., et al. (2013). A strategy to estimate unknown viral diversity in mammals. *mBio, 4*(5), Article e00598-00513. doi.org/10.1128/mBio.00598-13

22. Daszak, P., Amuasi, J., das Neves, C. G., et al. (2020). *IPBES (2020) Workshop Report on Biodiversity and Pandemics of the Intergovernmental Platform on Biodiversity and Ecosystem Services.* IPBES Secretariat. doi.org/10.5281/zenodo.4147317

23. Woolhouse, M., Scott, F., Hudson, Z., et al. (2012). Human viruses: Discovery and emergence. *Philosophical Transactions of the Royal Society B: Biological Sciences, 367*(1604), 2864–2871. doi.org/10.1098/rstb.2011.0354

24. Wolfe, N. D., Dunavan, C. P., & Diamond, J. (2007). Origins of major human infectious diseases. *Nature, 447*(7142), 279–283. doi.org/10.1038/nature05775

25. Johnson, C. K., Hitchens, P. L., Pandit, P. S., et al. (2020). Global shifts in mammalian population trends reveal key predictors of virus spillover risk. *Proceedings of the Royal Society B: Biological Sciences, 287*(1924), Article 20192736. doi.org/10.1098/rspb.2019.2736

26. Warren, C.J. & Sawyer, S.L. (2019). How host genetics dictates successful viral zoonosis. *PLoS Biology, 17*(4), e3000217. doi.org/10.1371/journal.pbio.3000217

27. Letko, M., Seifert, S. N., Olival, K. J., et al. (2020). Bat-borne virus diversity, spillover

and emergence. *Nature Reviews Microbiology, 18*(8), 461–471. doi.org/10.1038/s41579-020-0394-z

28. Moya, A., Elena, S. F., Bracho, A., et al. (2000). The evolution of RNA viruses: A population genetics view. *Proceedings of the National Academy of Sciences, 97*(13), 6967–6973. doi.org/10.1073/pnas.97.13.6967

29. Holmes, E. C., Dudas, G., Rambaut, A., et al. (2016). The evolution of Ebola virus: Insights from the 2013–2016 epidemic. *Nature, 538*(7624), 193–200. doi.org/10.1038/nature19790

30. Adam, D. (2020). A guide to *R* – the pandemic's misunderstood metric. *Nature, 583*, 346–348. doi.org/10.1038/d41586-020-02009-w

31. Guerra, F. M., Bolotin, S., Lim, G., et al. (2017). The basic reproduction number ($R_0$) of measles: A systematic review. *The Lancet Infectious Diseases, 17*(12), e420–e428. doi.org/10.1016/s1473-3099(17)30307-9

32. Kucharski, A. J., & Althaus, C. L. (2015). The role of superspreading in Middle East respiratory syndrome coronavirus (MERS-CoV) transmission. *Eurosurveillance, 20*(25), 14–18. doi.org/10.2807/1560-7917.es2015.20.25.21167

33. Chowell, G., Miller, M. A., & Viboud, C. (2008). Seasonal influenza in the United States, France, and Australia: Transmission and prospects for control. *Epidemiology & Infection, 136*(6), 852–864. doi.org/10.1017/S0950268807009144

34. Wong, Z. S., Bui, C. M., Chughtai, A. A., et al. (2017). A systematic review of early modelling studies of Ebola virus disease in West Africa. *Epidemiology & Infection, 145*(6), 1069–1094. doi.org/10.1017/S0950268817000164

35. Petersen, E., Koopmans, M., Go, U., et al. (2020). Comparing SARS-CoV-2 with SARS-CoV and influenza pandemics. *The Lancet Infectious Diseases, 20*(9), e238–e244. doi.org/10.1016/s1473-3099(20)30484-9

36. Locatelli, I., Trachsel, B., & Rousson, V. (2021). Estimating the basic reproduction number for COVID-19 in Western Europe. *PLoS ONE, 16*(3), Article e0248731. doi.org/10.1371/journal.pone.0248731

37. Liu, Y., & Rocklov, J. (2021). The reproductive number of the Delta variant of SARS-CoV-2 is far higher compared to the ancestral SARS-CoV-2 virus. *Journal of Travel Medicine, 28*(7), Article taab124. doi.org/10.1093/jtm/taab124

38. Richard, M., Knauf, S., Lawrence, P., et al. (2017). Factors determining human-to-human transmissibility of zoonotic pathogens via contact. *Current Opinion in Virology, 22*, 7–12. doi.org/10.1016/j.coviro.2016.11.004

39. Kraemer, M. U. G., Yang, C. H., Gutierrez, B., et al. (2020). The effect of human mobility and control measures on the COVID-19 epidemic in China. *Science, 368*(6490), 493–497. doi.org/10.1126/science.abb4218

40. Sartorius, B., Lawson, A. B., & Pullan, R. L. (2021). Modelling and predicting the spatio-temporal spread of COVID-19, associated deaths and impact of key risk factors

in England. *Scientific Reports, 11*(1), Article 5378. doi.org/10.1038/s41598-021-83780-2

41. Whitfield, J. T., Pako, W. H., Collinge, J., et al. (2008). Mortuary rites of the South Fore and kuru. *Philosophical Transactions of the Royal Society B: Biological Sciences, 363*(1510), 3721–3724. doi.org/10.1098/rstb.2008.0074

42. Brooks, J. (1996). The sad and tragic life of Typhoid Mary. *Canadian Medical Association Journal, 154*(6), 915–916. ncbi.nlm.nih.gov/pubmed/8634973

43. Woolhouse, M. E. J., Dye, C., Etard, J.-F., et al. (1997). Heterogeneities in the transmission of infectious agents: Implications for the design of control programs. *Proceedings of the National Academy of Sciences, 94*(1), 338–342. doi.org/10.1073/pnas.94.1.338

44. Galvani, A. P., & May, R. M. (2005). Epidemiology: Dimensions of superspreading. *Nature, 438*(7066), 293–295. doi.org/10.1038/438293a

45. Babkin, I. V., & Babkina, I. N. (2015). The origin of the variola virus. *Viruses, 7*(3), 1100–1112. doi.org/10.3390/v7031100

46. Roeder, P., Mariner, J., & Kock, R. (2013). Rinderpest: The veterinary perspective on eradication. *Philosophical Transactions of the Royal Society B: Biological Sciences, 368*(1623), Article 20120139. doi.org/10.1098/rstb.2012.0139

47. Baker, M. G., Wilson, N., & Blakely, T. (2020). Elimination could be the optimal response strategy for Covid-19 and other emerging pandemic diseases. *British Medical Journal, 2020*(371). doi.org/10.1136/bmj.m4907

48. Baker, M. G., Kvalsvig, A., & Verrall, A. J. (2020). New Zealand's COVID-19 elimination strategy. *The Medical Journal of Australia, 213*(5), 198–200, Article e191. doi.org/10.5694/mja2.50735

49. Geoghegan, J. L., Ren, X., Storey, M., et al. (2020). Genomic epidemiology reveals transmission patterns and dynamics of SARS-CoV-2 in Aotearoa New Zealand. *Nature Communications* 11(1), Article 6351. doi.org/ 10.1038/s41467-020-20235-8

50. Lee, A., Thornley, S., Morris, A. J., et al. (2020). Should countries aim for elimination in the Covid-19 pandemic? *British Medical Journal, 2020*(370), Article m3410. doi.org/10.1136/bmj.m3410

51. Wilson, N., Grout, L., Summers, J. A., et al. (2021). Use of the elimination strategy in response to the COVID-19 pandemic: Health and economic impacts for New Zealand relative to other OECD countries. *medRxiv preprint*. doi.org/10.1101/2021.06.25.2125 9556

52. Steyn, N., Binny, R. N., Hannah, K., et al. (2021). Māori and Pacific people in New Zealand have a higher risk of hospitalisation for COVID-19. *New Zealand Medical Journal, 134*(1538), 28–43.

53. Wilson, N., Mansoor, O. D., Boyd, M. J., et al. (2021). We should not dismiss the possibility of eradicating COVID-19: Comparisons with smallpox and polio. *BMJ Global Health, 2021*(6), Article e006810. doi.org/10.1136/bmjgh-2021-006810

54. Bootsma, M. C. J., & Ferguson, N. M. (2007). The effect of public health measures on the 1918 influenza pandemic in U.S. cities. *Proceedings of the National Academy of Sciences, 104*(18), 7588–7593. doi.org/10.1073/pnas.0611071104

55. Olliaro, P., Torreele, E., & Vaillant, M. (2021). COVID-19 vaccine efficacy and effectiveness – the elephant (not) in the room. *The Lancet Microbe, 2*(7), e279–e280. doi.org/10.1016/s2666-5247(21)00069-0

56. Dawood, F. S., Chung, J. R., Kim, S. S., et al. (2020). Interim estimates of 2019–20 seasonal influenza vaccine effectiveness – United States, February 2020. *Morbidity and Mortality Weekly Report, 69*(7), 177–182. doi.org/10.15585/mmwr.mm6907a1

57. Sanderson, K. (2021, August 19). COVID vaccines protect against Delta, but their effectiveness wanes. *Nature.* ncbi.nlm.nih.gov/pubmed/34413527

58. Morens, D. M., Taubenberger, J. K., & Fauci, A. S. (2022). Universal Coronavirus vaccines – An urgent need. *New England Journal of Medicine, 386*, 297–299. doi.org/10.1056/NEJMp2118468

59. Reichert, T. A., Sugaya, N., Fedson, D. S., et al. (2001). The Japanese experience with vaccinating schoolchildren against influenza. *New England Journal of Medicine, 344*(12), 889–896. doi.org/10.1056/NEJM200103223441204

60. Fine, P., Eames, K., & Heymann, D. L. (2011). 'Herd immunity': A rough guide. *Clinical Infectious Diseases, 52*(7), 911–916. doi.org/10.1093/cid/cir007

61. Adler, B. (2021, August 10). Vaccination responses paint worrying picture. *The Age.* theage.com.au/national/victoria/vaccination-responses-paint-worrying-picture-20210809-p58ha4.html

62. Health Act 1956. legislation.govt.nz/act/public/1956/0065/latest/whole.html

63. Civil Defence Emergency Management Act 2002. legislation.govt.nz/act/public/2002/0033/51.0/DLM149789.html

64. Epidemic Preparedness Act 2006. legislation.govt.nz/act/public/2006/0085/latest/whole.html

65. Ministry of Health. (2017). *New Zealand influenza pandemic plan: A framework for action* (2nd edn.). health.govt.nz/publication/new-zealand-influenza-pandemic-plan-framework-action

66. Roux, J., Granados, G. M., Shuey, L., et al. (2016). A unique genotype of the rust pathogen, *Puccinia psidii*, on Myrtaceae in South Africa. *Australasian Plant Pathology, 45*(6), 645–652. doi.org/10.1007/s13313-016-0447-y

67. Dayton, L., & Higgins, E. (2011, April 9). Myrtle rust 'biggest threat to ecosystem'. *The Australian.* theaustralian.com.au/news/health-science/myrtle-rust-biggest-threat-to-ecosystem/news-story/38cfaa914c6617476382f9b5b14d573c

68. Wilfert, L., Long, G., Leggett, H. C., et al. (2016). Deformed wing virus is a recent global epidemic in honeybees driven by *Varroa* mites. *Science, 351*(6273), 594–597. doi.org/10.1126/science.aac9976

69. Traynor, K. S., Mondet, F., de Miranda, J. R., et al. (2020). *Varroa destructor*: A complex parasite, crippling honey bees worldwide. *Trends in Parasitology, 36*(7), 592–606. doi.org/10.1016/j.pt.2020.04.004

70. Natsopoulou, M. E., McMahon, D. P., Doublet, V., et al. (2017). The virulent, emerging genotype B of *Deformed wing virus* is closely linked to overwinter honeybee worker loss. *Scientific Reports, 7*(1), Article 5242. doi.org/10.1038/s41598-017-05596-3

71. Messenger, A. M., Barnes, A. N., & Gray, G. C. (2014). Reverse zoonotic disease transmission (zooanthroponosis): A systematic review of seldom-documented human biological threats to animals. *PLoS ONE, 9*(2), Article e89055. doi.org/10.1371/journal.pone.0089055

72. Cerda-Cuellar, M., More, E., Ayats, T., et al. (2019). Do humans spread zoonotic enteric bacteria in Antarctica? *Science of the Total Environment, 654*, 190–196. doi.org/10.1016/j.scitotenv.2018.10.272

73. Sooksawasdi Na Ayudhya, S., & Kuiken, T. (2021). Reverse zoonosis of COVID-19: Lessons from the 2009 influenza pandemic. *Veterinary Pathology, 58*(2), 234–242. doi.org/10.1177/0300985820979843

74. USDA-APHIS. (2021). *Questions and answers: Results of study on SARS-CoV-2 in white tailed deer.* aphis.usda.gov/animal_health/one_health/downloads/qa-covid white-tailed-deer-study.pdf

75. Hydatids Act 1959. nzlii.org/nz/legis/hist_act/ha19591959n93130.pdf

76. Davidson, R. M. (2002). Control and eradication of animal diseases in New Zealand. *New Zealand Veterinary Journal, 50*(3 Suppl), 6–12. doi.org/10.1080/00480169.2002.36259

77. Marani, M., Katul, G. G., Pan, W. K., et al. (2021). Intensity and frequency of extreme novel epidemics. *Proceedings of the National Academy of Sciences, 118*(35), Article e2105482118. doi.org/10.1073/pnas.2105482118

78. Dobson, A. P., Pimm, S. L., Hannah, L., et al. (2020). Ecology and economics for pandemic prevention. *Science, 369*(6502), 379–381. doi.org/10.1126/science.abc3189

79. McKnight, M. (2020, May 27). The virus hunter: Into the wild. Twice. For mankind. *Sports Illustrated.*

80. Jordan, D., Tumpey, T. & Jester, B. (2019). *The deadliest flu: The complete story of the discovery and reconstruction of the 1918 pandemic virus.* Centers for Disease Control and Prevention, National Center for Immunization and Respiratory Diseases.

81. Tumpey, T.M. et al. (2005). Characterization of the reconstructed 1918 Spanish influenza pandemic virus. *Science, 310*(5745), 77–80. doi.org/10.1126/science.1119392

82. Lytras, S., Xia, W., Hughes, J., Jiang, X. & Robertson, D.L. (2021). The animal origin of SARS-CoV-2. *Science, 373*(6558), 968–970. doi.org/10.1126/science.abh0117

83. Song, H.D. et al. (2005). Cross-host evolution of severe acute respiratory syndrome

coronavirus in palm civet and human. *Proceedings of the National Academy of Sciences, 102* (7), 2430–2435. doi.org/10.1073/pnas.0409608102

84. Maxmen, A. & Mallapaty, S. (2021, June 17). The COVID lab-leak hypothesis: What scientists do and don't know. *Nature, 594*(7863), 313–315. doi.org/10.1038/d41586-021-01529-3

85. Xiao, X., Newman, C., Buesching, C.D., Macdonald, D.W. & Zhou, Z.M. (2021). Animal sales from Wuhan wet markets immediately prior to the COVID-19 pandemic. *Scientific Reports, 11(*11898). doi.org/10.1038/s41598-021-91470-2

86. Conceicao, C. et al. (2020). The SARS-CoV-2 Spike protein has a broad tropism for mammalian ACE2 proteins. *PLoS Biology, 18,* e3001016. doi.org/10.1371/journal.pbio.3001016

87. Pinto, B.G.G. et al. (2020). ACE2 expression is increased in the lungs of patients with comorbidities associated with severe COVID-19. *Journal of Infectious Disease, 222*(4), 556–563. doi.org/10.1093/infdis/jiaa332

88. Callaway, E. (2021). The mutation that helps Delta spread like wildfire. *Nature, 596*(7873), 472–473. doi.org/10.1038/d41586-021-02275-2

# 7. THE FUTURE

## Even more pathogens, pests and pandemics

So, where are we at in our world of pests and pestilence? Here is a quick summary.

First, our problems with invasive species, pests and pathogens are bad and are going to get worse. Globalisation and trade have brought us wonderful things but also a plethora of new pests and pathogens. For most taxa groups, the rate of pests being intercepted for the first time in an area ('first record') has been increasing, with highest rates observed in recent years.[1] Newly arriving pests and invasive species join communities of unintentionally introduced rats, mice and others, including wilding pines and the plant pathogen myrtle rust. Pandemics are a subset of our pest problems, but they can bring the world to its knees. We can expect more pandemics and more pest problems here in New Zealand and around the globe, and similarly with pathogens and their vectors. As Tatem, Rogers and Hay (2006) wrote, 'With no apparent end in sight to the continued growth in global air travel and shipborne trade, we must expect the continued

Cane toads introduced to Australia have evolved longer legs, a change associated with increased dispersal and invasion rates. They have also developed a propensity for cannibalism not seen in their native range. *Buiten-Beeld / Alamy*

appearance of communicable disease pandemics, disease-vector invasions and vector-borne disease movement.'[2]

Second, the increasing problems we are experiencing with pests and pestilence are tightly intertwined with population growth, globalisation and international movement, and anthropogenic-driven change that includes habitat destruction, biodiversity exploitation and climate change (Fig. 7.1). How we interact with and move animals, plants and pathogens is at the core of these problems. The emergence and distribution of many major pandemics involve zoonoses and pests, such as the bubonic plague pandemics that involved humans moving

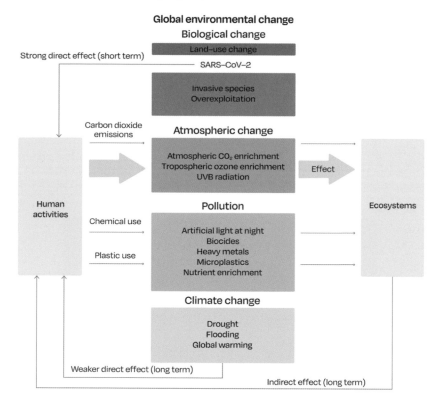

Fig. 7.1: Many scientists believe that the COVID-19 pandemic is a symptom or part of a syndrome of human effects on the planet; that pandemic and environmental change are intimately intertwined, with feedback loops. The emergence of the SARS-CoV-2 pathogen in humans is probably the result of human–wildlife contact. Its global spread in many ways resembles the process of invasive species spread. Global cooperation on environmental change, invasive species and health is needed. *Adapted from Rillig et al. (2021)[34]*

invasive species (rats), their parasites (fleas) and the bacterial disease (*Yersinia pestis*) that they carry. We can sometimes eradicate or control invasive species and pathogens, but these controls are not always effective and have issues that include cost and social acceptance. New control approaches are needed, especially if we are to achieve the lofty goals of the Predator Free 2050 campaign, which aims to eradicate a range of predators from New Zealand.

A worrying aspect of these invasive species, pests and pathogens is the rate at which they can evolve and change. With pathogens, we are all too familiar with how the SARS-CoV-2 virus evolved into the Delta variant, which is estimated to be at least 40% more transmissible than the ancestral virus.[3] The Omicron variant has proved to be even more infectious. Many vertebrate and invertebrate pests also seem capable of rapid evolution. Invasive house mice (*Mus musculus*) on remote Gough Island rapidly evolved to become larger, and to work in groups to attack and kill otherwise healthy albatross chicks that were 300 times their mass.[4] Similarly, after just 70 years in Australia, cane toads (*Rhinella marina*) showed rapid adaptive change by evolving longer legs that increased their rate of dispersal.[5] These toads have evolved other unusual behaviours in their newly introduced range, including an increased propensity to cannibalise younger toads, as well as behavioural and physiological mechanisms to avoid cannibalism.[6] In New Zealand, pest species have also shown an extraordinarily

House mice on Gough Island have become larger, attacking and killing otherwise healthy chicks. Here a mouse feeds from a wound it has caused on an endangered Tristan albatross (*Diomedea dabbenena*).
*Ben Dilley photo*

fast rate of adaptation and evolution. An important pasture pest here is the Argentine stem weevil (*Listronotus bonariensis*). Over just 21 years, this weevil evolved resistance to what was once an effective biological control agent.[7]

Finally, the international treaties and national frameworks for dealing with pests and pestilence have come a long way since James Cook happily watched rats scuttle off his ship in the Pacific Islands. But have our attempts at legislation or regulation been effective? Has the Convention on Biological Diversity, for example, been successful? Let's just conclude that it could have been much, much more effective. Even New Zealand's national legislation for dealing with pests could use improvement. Our national and regional pest management plans are failing for many pests, including rabbits and wallabies, while biosecurity in our marine ecosystems is sorely lacking. On a similar theme, for pandemic management, a key tool for the WHO has been to implement an obligation for Public Health Emergency of International Concern notices. Nevertheless, here we are with COVID-19 rampant in every corner of the globe, and the official death toll at 6 million in 2022. We have vaccines and other tools to deal with the SARS-CoV-2 virus, but with better proactive tools perhaps we could have avoided it altogether, or have found and controlled it before it become a pandemic.

We need effective biosecurity and pest control. We need new tools and approaches. How do we get there? What are the emerging options?

## New approaches: One Health and One Biosecurity

The global eradication of rinderpest was outstanding. Veterinary and wildlife ecologists around the world worked with socio-ecological approaches to disease investigation and treatment.[8] It was a multidisciplinary approach to a highly contagious, major global disease affecting wildlife and livestock.

One Health largely extends the approach taken against rinderpest, but with a focus on human health. A definition of One Health is 'the collaborative effort of multiple health science professions, together with their related disciplines and institutions – working locally, nationally and globally – to attain optimal health for people, domestic animals, wildlife, plants and our environment'.[9] This concept developed from the observation that medical and veterinary science share common interests in many diseases, and that the understanding of wildlife plays a central role in understanding disease emergence in humans (Fig. 7.2).

The poster child for One Health has been the response to the H1N1 influenza pandemic that emerged in 2009. A key component was a surveillance framework involving several organisations, including the WHO, the Food and Agriculture Organization of the United Nations, the World Organization for Animal Health, the World Bank and others. This surveillance system now detects new and evolving avian influenza strains that have the potential to cause widespread human disease.[9]

The COVID-19 pandemic has devastatingly demonstrated how zoonotic diseases can be quickly transported to every corner of the world (there have even been cases in Antarctica), and then back to animals in a reverse zoonosis. The SARS-CoV-2 virus has beaten the best biosecurity systems currently in use. Our understanding and control of this pandemic will require the collaboration

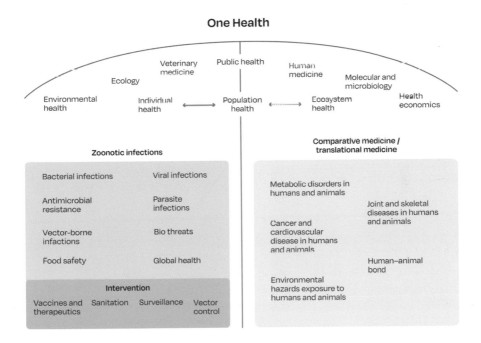

Fig. 7.2: The 'One Health Umbrella' was developed by the networks One Health Initiative and One Health Sweden. The concept of One Health is described as a strategy for expanding interdisciplinary collaborations and communications in all aspects of healthcare for humans, animals and the environment. The aim is to advance healthcare by accelerating biomedical research discoveries, enhancing public health efficacy, expanding the scientific knowledge base, and improving medical education and clinical care. *Adapted from onehealthinitiative.com*

of scientists from a range of fields, from ecology and virology to genomics, and health professionals and governments around the world. The One Health initiative may have prevented this pandemic if it had been widely implemented.

Perhaps, however, even One Health is too limited in its scope. Some of the world's worst invasive species have a multitude of deleterious effects, such as the giant African snail (*Lissachatina fulica*) that simultaneously destroys crops and spreads both plant and human diseases. Another of the world's worst pest is the mosquito, the pest that has cost us the most money by far for management and in damage between 1970 and 2017 (Fig. 1.4). In 2020, Philip Hulme from Lincoln University in New Zealand suggested taking One Health a step further: One Biosecurity would be an interdisciplinary and unified concept to build 'on the interconnections between human, animal, plant, and environmental health to effectively prevent and mitigate the impacts of invasive alien species' (Fig. 7.3).[10] He suggests three initiatives to effectively manage pandemic risks associated with biological invasions and pests; first, that new surveillance and risk-assessment

Left: A boy holds a giant African snail in South East Cameroon. Right: A curious baboon inspects a snail. In China and other countries, the snail has been responsible for outbreaks of Eosinophilic meningitis in humans and livestock caused by the transmission of rat lungworm (*Angiostrongylus cantonensis*), which the snail hosts.[35] The snail can damage crops directly, both by feeding and transmitting plant pathogens (*Phytophthora* spp.).[36] This snail is an example of a species with a multitude of impacts across the environment, biodiversity and human or livestock health. *Left: Nature Picture Library / Alamy. Right: B.G. Wilson Wildlife / Alamy*

tools are needed to look beyond national borders toward biosecurity risks of international concern. Second, a stronger regulatory framework or instrument is necessary to address biosecurity threats at a worldwide scale. Third, to implement such a framework, a multilateral biosecurity convention responsible for biosecurity governance is required.[11] These concepts are well supported. Montserrat Vilà et al. (2021) conclude that:

> Fundamental concepts in invasion biology regarding the interplay of propagule pressure [the rate of arrival of a non-native species in a country or site, in terms of the number arriving in each group and the number of groups], species traits [such as the rate of reproduction, their diet, etc], biotic interactions [such as competition or predation], eco-evolutionary experience, and ecosystem disturbances can help to explain transitions between stages of epidemic spread. As a result, many forecasting

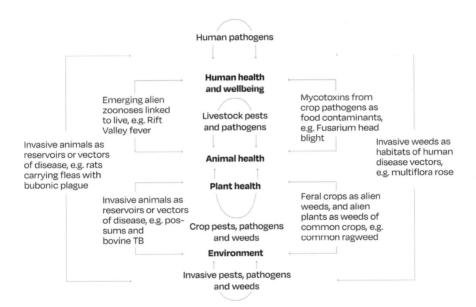

Fig. 7.3: A schematic representation of the One Biosecurity concept, emphasising the interwoven nature of human, animal, plant and environmental health that arises through the impacts of invasive plants, animals and pathogens. For example, invasive rats can carry fleas and zoonotic diseases including bubonic plague, which can affect wildlife and human populations. As predators of birds, rats can alter patterns of seed dispersal, which can change plant communities. *Adapted from Hulme (2020)[10]*

and management tools used to address epidemics could be applied to biological invasions and vice versa. Therefore, we advocate for increasing cross-fertilization between the two disciplines to improve prediction, prevention, treatment, and mitigation of invasive species and infectious disease outbreaks, including pandemics.[12]

There is considerable overlap between many invasive pests, diseases, and human health and economic well-being.

I'm not sure what the appetite is for the new conventions and global cooperation that would be required for One Biosecurity, especially in countries that are central to our need for surveillance of zoonotic diseases and pests, such as China. It is clear, however, that new approaches are needed for us to cope with an increasing number of pest interceptions and pandemics. The surveillance and monitoring to prevent zoonotic disease spillover is a key strategy against pandemic development.[13] The cost–benefit analyses discussed in the previous chapter highlight just how beneficial surveillance and proactive measures can be. As a reminder, the estimated global total gross prevention costs for pandemics would range between US$22 and 31 billion. The programmes would be for detection and control, reducing spillover from livestock, reducing deforestation and ending the wild meat trade in China. If such programmes had been in place, we may well have prevented the COVID-19 pandemic, which in 2020 was estimated to have cost between US$8.1 and 15.8 trillion.[14]

## Current and future tools for pest control

Several particularly problematic pests have received intensive research over many decades, leading to an array of approaches for current attempts at their control but also the acknowledgement that new tools and innovative ideas are desperately needed for their management into the future. One exemplar problematic pest is the cane toad, which was intentionally introduced into countries such as Australia for biological control of insect pests. Largely due to their poison glands, cane toads have been devastating for native biodiversity in Australia. Freshwater crocodiles, lizards, snakes and other large anuran-eating predators have had their populations crash, after being fatally poisoned by eating the toads.[15] These declines have led trophic cascades that have reshaped ecological communities.[16]

Australian scientists have put substantial resources into understanding and controlling cane toads.

The current methods used for suppressing the population of cane toads include many of the familiar and traditional approaches.[15] Toads have been manually captured, trapped, shot, killed and removed, and they have been fenced away from water resources. Wildlife managers have considered how to use pathogens or native predators as biological control agents, although biological control via pathogens has largely been abandoned because of concerns about the potential effects on non-target species. Communities desperate for toad control have used pesticides. The use of these chemicals, however, has had adverse ecological impacts and has typically been abandoned. A search on YouTube will show you some less familiar management approaches. You'll see Australians gleefully implementing toad control with vans swerving madly over highways to flatten as many as they can. In 2005 Australian member of parliament Dave Tollner controversially recommended a 'whacking day', whereby the public were urged to bludgeon cane toads to death with golf clubs or cricket bats. Dave had grown up using these toad-control techniques and believed that the welfare rights of

Most species of Australian snake die after trying to eat cane toads, as the toads secrete potent cardiac toxins. After this interaction shown above, the carpet python (*Morelia spilota*) released the toad. 'The cane toad was pretty pissed off with the snake,' the photographer remarked, 'and decided to get back at the snake and attacked him back. Bad decision. There was no second chance for the toad and he met his demise. The snake did not eat the toad. Sad thing is we found the carpet snake dead a month later with a cane toad inside him.' *Barbara Jean / Alamy*

native animals had to be considered before the rights of the cane toad. 'I mean a cane toad can cause a slow death in a crocodile or a goanna or any other animal that eats it,' Tollner said. 'I think at times we overlook our native wildlife and at times certain organisations can merely look at what's best for the toad.'[17] Animal-welfare groups were horrified, and suggested the best way to kill the toads was to collect them in plastic bags and freeze them.

Next-generation technologies and approaches are being considered for cane toad suppression.[15] Larval toads are known to produce a pheromone that interferes with embryonic development of younger conspecifics. These 'suppression pheromones' could be synthesised and sprayed into the environment, although it seems unlikely to me that they would be sprayed in large enough spatial scales and with the frequency needed for widespread toad control. Monitoring approaches have been developed using automated toad-calling detectors, and machines that mimic toad calls have been produced to attract toads for their control. In order to reduce the rate of toad spread, scientists are considering 'genetic backburning' or seeding toads ahead of the hyper-dispersive genotype that is spearheading the toad invasion. The presence of a 'regular' toad genotype in a region before the hyper-dispersive genotype arrives would result in interbreeding between the cane toad genotypes, with the resulting offspring being slower dispersers. Behavioural researchers have developed 'taste-aversion training' approaches to deter native Australian predators from eating cane toads. Geneticists have also developed gene-editing technologies that could be used to disable the toads' toxin production. Other gene-drive targets could be to skew the sex ratio, perhaps using a form of genetic immunocontraception so that only males are produced. The Australian branch of the animal rights organisation People for the Ethical Treatment of Animals support toad management by genetic modification that would induce immunocontraception. They see this as a 'humane, long-term population-control technique'.[18]

Outside the world of cane toads, scientists recognise that other sectors are similarly at a critical juncture in pest management. Weed scientists, for example, are concerned about the development of pesticide resistance in weed pests and the need to feed nearly 10 billion people by 2050. They recognise a need to develop new herbicides and safer biopesticides, and the potential of genetic control methods, including gene silencing or RNAi pesticides and gene-drive technologies. These tools could be used in combination with current approaches such as biological control.[19]

Those involved with mosquito or disease-vector control have had success with disease-control methods such as the release of *Wolbachia*-infected mosquitoes to control dengue fever.[20] But substantial challenges in this field remain. In a 2017 review, Giovanni Benelli and John Beier suggested that 'current strategies for malaria vector control used in most African countries are not sufficient to achieve successful malaria control'.[21]

It is likely that there will be no 'silver bullet' for the control of cane toads or any other pest species. Achieving suppression or eradication will require different tools for different pests. Societal acceptance of the pest management approaches, or a social licence to operate, is often a limiting factor in the implementation of many of the next-generation technologies. Within these discussions about different approaches, it will be important to consider not only the question 'What is the cost of trying?' but also 'What is the cost of doing nothing?'[15]

## Pest management should be ethical

The management of cane toads in Australia typically involves killing individuals, whether with a golf club or more humane means. But should we kill in the name of conservation or pest control? Programmes such as the Predator Free 2050 campaign in New Zealand will likely involve the killing of many millions of rats, mustelids and possums, all of which are considered sentient, meaning capable of perceiving and feeling things. What is our moral authority for this management action? Conservation biologists are increasingly being asked these questions.

'Compassionate conservationists' argue that sentient animals are effectively persons and should have a similar moral status to human beings. Many compassionate conservationists seek to completely 'avoid deliberately harming sentient beings' in conservation programmes.[22] They argue that no killing for conservation is justified[23] and that many existing conservation programmes are morally wrong. Programmes in countries such as New Zealand typically address the leading cause of population declines with a focus on killing or inhibiting the free movement of vertebrate predators. Such programmes run completely counter to the ethos of compassionate conservation.

There have been some very questionable methods used in pest control programmes in the name of conservation. In Australia, another conservation programme sought to control a population of introduced wild goats (*Capra hircus*) on Pelorus Island. The programme would translocate captured mainland dingoes (*Canis dingo*) onto the island, which would eradicate the goats and then be shot. Male dingoes were trapped and reproductively sterilised. In case they were not able to be found and shot, the dingoes were implanted with poison capsules timed to kill them within two years. Many conservation practitioners had problems with this approach, and public protest ended the programme after two dingoes were put on the island.[24]

The other side of the coin is that doing nothing can do considerable harm. Sentient beings are clearly being harmed by invasive species in the cases of invasive mice attacking albatross chicks on subantarctic islands, cane toads killing snakes, or possums and rats eating bird chicks in New Zealand. Animal-welfare goals may be best achieved by limiting population sizes or entirely eradicating invasive pests. Introduced herbivores that become overly abundant, for example, can starve and impact other species through resource depletion and competition.[25] Introduced predators exert a clear and considerable animal-welfare impact. Introduced mammalian predators of kiwi in New Zealand kill at least 8% of chicks, 45% of juveniles and possibly as many as 60% of young kiwi.[26]

I also suspect that many compassionate conservationists don't see or directly interact with the consequences of pest plagues. They don't watch defenceless albatross chicks being slowly eaten alive. They don't reside in or have families living in areas with pest plagues. John Southon, principal at Trundle Central School in central west New South Wales, described such a life:

> Nobody understands a mouse plague until you've lived through it. Nobody understands the absolutely pungent smell, the fact that your furniture is eaten, it's just horrendous. The mice have eaten all the insulation in our air conditioning systems. They've eaten wires out of the roof of the school, they've eaten parts of the power board in the principal's residence. It's just constant. To me, I would describe it as having an injury where you're just in constant pain. Eventually, it's going to affect your mental health.[27]

How do we go about ethical pest control for conservation? In 2015 a workshop was convened with the goal of developing the first international principles for ethical decision-making in wildlife control. From this workshop, experts from the conservation industry, academia and nongovernmental organisations in five continents went on to publish a series of recommendations for the ethical control of pests.[28] In a 2017 article, Sara Dubois et al. recommend that any effort to control pest populations should:

1.  begin by determining whether the problem could be mitigated by changing the behaviour of people, developing a culture of coexistence;
2.  be justified by evidence that significant or serious harm is being caused by the target pest species;
3.  have clear, outcome-based and measurable objectives that are achievable, monitored and adaptive;
4.  ensure the pest control method causes the least possible harm to the fewest possible animals;
5.  be informed by community values as well as scientific, technical and practical information;
6.  be integrated into plans for systematic long-term management; and
7.  be based on the specifics of the situation rather than negative labels applied to the target species.[28]

These recommendations for the ethical control of pests have been applied to the badger culling in England, used as a means to control bovine tuberculosis (*Mycobacterium bovis*), with a former UK deputy chief veterinary officer finding no support for continuing the cull.[29] For many other species considered to be pests, in contrast, I'm confident that the vast majority of people would find many control operations to be highly ethical under this framework. For example, the use of toxic bait for the eradication of house mice (*Mus musculus*) on Gough Island will be immensely beneficial for the welfare and survival of many native birds, including the albatross.

Many pest managers and conservation biologists reject the 'compassionate conservationist' argument, recognising the potential for grave biodiversity loss if it were to be implemented. They recognise that finding the 'least bad' option is difficult but often necessary.[23] I think the criteria developed by Sara Dubois and colleagues might be a step in that direction.

## Rapid and comprehensive action is needed

We have the potential to substantially reduce the future impacts of pests and pestilence, but rapid and comprehensive action is needed.[30] Pest and pestilence management will be increasingly important over the coming decades. The recommendations by Goldson et al. in a 2015 publication on pest management in New Zealand echoed these conclusions:

> This contribution has sought to highlight the inescapable importance of effective pest management in New Zealand under changing circumstances. Considerations include accelerating globalisation, the ongoing arrival of invasive species, declining effectiveness and acceptability of existing controls, intensifying land use and climate change. Accentuating this is the growing demand by international trading partners for quality assurance. Doing nothing about these trends and conditions is not an option . . . The inherent limitations of existing pest management approaches underline the requirement for discovery, new technologies and ongoing refinement of existing methods. In particular, there needs to be an overall move away from the use of pesticides to control systems based on good biological understanding. Such a shift would extend beyond the study of individual organisms to ecosystem function and susceptibility to invasion by exotic species. All of this calls for a commensurate increase in expertise. With this, it will be necessary to engage early with the public over novel pest control tools and strategies, or risk losing the battle for pest control.[31]

In 2021, my family visited New Zealand's Northland region and included a visit to Tāne Mahuta ('Lord of the Forest'). Tāne Mahuta is an awe-inspiring kauri tree (*Agathis australis*) that may be 2500 years old. It stands at 50 metres tall, with a girth of nearly 14 metres. The tree reportedly has 47 different epiphyte plant species growing in its crown.[32] In New Zealand, kauri and these epiphytes offer a habitat for an impressive range of invertebrates, many of which are undescribed and 'new' to science. Māori regard Tāne Mahuta as a living ancestor.

Unfortunately, Tāne Mahuta is threatened by kauri dieback, a disease caused by the oomycete *Phytophthora agathidicida* that kills kauri and, worryingly, has been reported within 60 metres of Tāne Mahuta.[33] Solutions for kauri dieback are desperately needed for all kauri. While gazing at this tree, it was hard not to wonder if future generations will have the same privilege.

We need urgent action for the long-term conservation of this and many other treasured species. We need effective pest control for our economies and to feed the estimated nearly 10 billion people that will be on the planet by 2050. The health of these people is also in our generations' hands, in the way we deal with pandemics and diseases that are so intrinsically related to our management of biodiversity.

Tāne Mahuta is the largest known living kauri (*Agathis australis*) in New Zealand and may be 2500 years old. The tree is threatened by kauri dieback, a relative of the disease that causes potato late blight and that caused the Irish potato famine. This disease or pest has killed entire stands of kauri. Solutions for kauri dieback are desperately needed: I hope my children's children and many future generations in New Zealand live with Tāne Mahuta. *Phil Lester photo*

# References

1. Seebens, H., Blackburn, T. M., Dyer, E. E., et al. (2017). No saturation in the accumulation of alien species worldwide. *Nature Communications, 8*, Article 14435. doi.org/10.1038/ncomms14435

2. Tatem, A. J., Rogers, D. J., & Hay, S. I. (2006). Global transport networks and infectious disease spread. *Advances in Parasitology, 62*, 293–343. doi.org/10.1016/S0065-308X(05)62009-X

3. Callaway, E. (2021). The mutation that helps Delta spread like wildfire. *Nature, 596*(7873), 472–473. doi.org/10.1038/d41586-021-02275-2

4. Wanless, R. M., Angel, A., Cuthbert, R. J., et al. (2007). Can predation by invasive mice drive seabird extinctions? *Biology Letters, 3*(3), 241–244. doi.org/10.1098/rsbl.2007.0120

5. Phillips, B. L., Brown, G. P., Webb, J. K., et al. (2006). Invasion and the evolution of speed in toads. *Nature, 439*, 803. doi.org/10.1038/439803a

6. DeVore, J. L., Crossland, M. R., Shine, R., et al. (2021). The evolution of targeted cannibalism and cannibal-induced defenses in invasive populations of cane toads. *Proceedings of the National Academy of Sciences, 118*(35), Article e2100765118. doi.org/10.1073/pnas.2100765118

7. Tomasetto, F., Tylianakis, J. M., Reale, M., et al. (2017). Intensified agriculture favors evolved resistance to biological control. *Proceedings of the National Academy of Sciences, 114*(15), 3885–3890. doi.org/10.1073/pnas.1618416114

8. Roeder, P., Mariner, J., & Kock, R. (2013). Rinderpest: The veterinary perspective on eradication. *Philosophical Transactions of the Royal Society B: Biological Sciences, 368*(1623), Article 20120139. doi.org/10.1098/rstb.2012.0139

9. Gibbs, E. P. J. (2014). The evolution of One Health: A decade of progress and challenges for the future. *Veterinary Record, 174*(4), 85–91. doi.org/10.1136/vr.g143

10. Hulme, P. E. (2020). One Biosecurity: A unified concept to integrate human, animal, plant, and environmental health. *Emerging Topics in Life Sciences, 4*(5), 539–549. doi.org/10.1042/ETLS20200067

11. Hulme, P. E. (2021). Advancing One Biosecurity to address the pandemic risks of biological invasions. *BioScience, 71*(7), 708–721. doi.org/10.1093/biosci/biab019

12. Vilà, M., Dunn, A. M., Essl, F., et al. (2021). Viewing emerging human infectious epidemics through the lens of invasion biology. *BioScience, 71*(7), 722–740. doi.org/10.1093/biosci/biab047

13. Vora, N. M., Sizer, N., & Bernstein, A. (2021). Preventing spillover as a key strategy against pandemics. *Nature, 597*(7876), 332. doi.org/10.1038/d41586-021-02427-4

14. Dobson, A. P., Pimm, S. L., Hannah, L., et al. (2020). Ecology and economics for pandemic prevention. *Science, 369*(6502), 379–381. doi.org/10.1126/science.abc3189

15. Tingley, R., Ward-Fear, G., Schwarzkopf, L., et al. (2017). New weapons in the toad toolkit: A review of methods to control and mitigate the biodiversity impacts of invasive cane toads (*Rhinella Marina*). *The Quarterly Review of Biology, 92*(2), 123–149. doi.org/10.1086/692167

16. Doody, J. S., Castellano, C. M., Rhind, D., et al. (2013). Indirect facilitation of a native mesopredator by an invasive species: Are cane toads re-shaping tropical riparian communities? *Biological Invasions, 15*(3), 559–568. doi.org/10.1007/s10530-012-0308-8

17. AAP. (2005, April 11). 'Whacking day: MP urges public to kill cane toads'. *The Sydney Morning Herald.* smh.com.au/national/whacking-day-mp-urges-public-to-kill-cane-toads-20050411-gdl3ub.html

18. PETA Australia. (2021). *Cane toad hunting: Why cruelty to animals is never the answer.* peta.org.au/issues/entertainment/cane-toad-hunting-cruelty-animals-never-answer/

19. Westwood, J. H., Charudattan, R., Duke, S. O., et al. (2018). Weed management in 2050: Perspectives on the future of weed science. *Weed Science, 66*(3), 275–285. doi.org/10.1017/wsc.2017.78

20. Utarini, A., Indriani, C., Ahmad, R. A., et al. (2021). Efficacy of *Wolbachia*-infected mosquito deployments for the control of dengue. *The New England Journal of Medicine, 384*(23), 2177–2186. doi.org/10.1056/NEJMoa2030243

21. Benelli, G., & Beier, J. C. (2017). Current vector control challenges in the fight against malaria. *Acta Tropica, 174*, 91–96. doi.org/10.1016/j.actatropica.2017.06.028

22. Wallach, A. D., Batavia, C., Bekoff, M., et al. (2020). Recognizing animal personhood in compassionate conservation. *Conservation Biology, 34*(5), 1097–1106. doi.org/10.1111/cobi.13494

23. Johnson, P. J., Adams, V. M., Armstrong, D. P., et al. (2019). Consequences matter: Compassion in conservation means caring for individuals, populations and species. *Animals, 9*(12), 1115. doi.org/10.3390/ani9121115

24. Wallach, A. D., Bekoff, M., Batavia, C., et al. (2018). Summoning compassion to address the challenges of conservation. *Conservation Biology, 32*(6), 1255–1265. doi.org/10.1111/cobi.13126

25. Hampton, J. O., Warburton, B., & Sandoe, P. (2019). Compassionate versus consequentialist conservation. *Conservation Biology, 33*(4), 751–759. doi.org/10.1111/cobi.13249

26. McLennan, J. A., Potter, M. A., Robertson, H. A., et al. (1996). Role of predation in the decline of kiwi, *Apteryx* spp, in New Zealand. *New Zealand Journal of Ecology, 20*(1), 27–35. https://newzealandecology.org/nzje/1986

27. Boseley, M. (2021, May 14). Australia's mouse plague: Six months ago it was war, now whole towns have accepted their presence. *The Guardian*. theguardian.com/australia-news/2021/may/15/australias-mouse-plague-six-months-ago-it-was-war-now-whole-towns-have-accepted-their-presence#comments

28. Dubois, S., Fenwick, N., Ryan, E. A., et al. (2017). International consensus principles for ethical wildlife control. *Conservation Biology, 31*(4), 753–760. doi.org/10.1111/cobi.12896

29. Simmons, A. (2020). Killing badgers to control bTB is unethical. *Vet Record, 186*(11), 357–358. doi.org/10.1136/vr.m1131

30. Essl, F., Lenzner, B., Bacher, S., et al. (2020). Drivers of future alien species impacts: An expert-based assessment. *Global Change Biology, 26*(9), 4880–4893. doi.org/10.1111/gcb.15199

31. Goldson, S. L., Bourdot, G. W., Brockerhoff, E. G., et al. (2015). New Zealand pest management: Current and future challenges. *Journal of the Royal Society of New Zealand, 45*(1), 31–58. doi.org/10.1080/03036758.2014.1000343

32. Warne, K. (2014, Jul–Aug). The future of our forests. *New Zealand Geographic*. nzgeo.com/stories/the-future-of-our-forests/

33. Roy, E. A. (2018, July 14). 'Like losing family': Time may be running out for New Zealand's most sacred tree. *The Guardian*. theguardian.com/world/2018/jul/14/like-losing-family-time-may-be-running-out-for-new-zealands-most-sacred-tree

34. Rillig, M.C., Lehmann, A., Bank, M. S., Gould, K. A. & Heekeren, H. R. (2021). Scientists need to better communicate the links between pandemics and global environmental change. *Nature Ecology & Evolution, 5*(11), 1466–1467. doi.org/10.1038/s41559-021-01552-7

35. Lv, S., Zhang, Y., Liu., H.-Z., et al. (2009). Invasive snails and an emerging infectious disease: results from the first national survey on *Angiostrongylus cantonensis* in China. *PLOS Neglected Tropical Diseases, 3*, e368. doi.org/10.1371/journal.pntd.0000368

36. Kant, R. & Diarra, S. S. (2016). Feeding strategies of the giant African snail *Achatina fulica* on papaya in Samoa. *ISHS Acta Horticulturae, 1128*, 229–235. doi.org/10.17660/ActaHortic.2016.1128.35

# Acknowledgements

First, a big vote of thanks to Ashleigh Young from Te Herenga Waka University Press for your help with this and my previous books. Theresa Crewdson also helped with editing and substantially improved the book. Many thanks to everyone who read and commented on draft chapters. I've been fortunate to have everyone from family, law professors, statisticians, and graduate students taking my class offer valued comments and suggestions. These kind people have included James Baty, Mariana Bulgarella, Eric Edwards, Antoine Felden, Susy Frankel, John Haywood, Matt Howse, Sarah Lester, Winifred Long, Rose McGruddy, Winifred Long, Tessa Pilkington, Aiden Reason, Zoe Smeele and many others.

# Index